William Watson

The Medical Adviser

A Complete Treatise on the Formation, Debility and Diseases of those

Organs Peculiar in Each Sex. Tenth Edition

William Watson

The Medical Adviser
*A Complete Treatise on the Formation, Debility and Diseases of those Organs
Peculiar in Each Sex. Tenth Edition*

ISBN/EAN: 9783337790714

Printed in Europe, USA, Canada, Australia, Japan

Cover: Foto ©berggeist007 / pixelio.de

More available books at **www.hansebooks.com**

THE MEDICAL ADVISER:

A COMPLETE TREATISE

ON THE

FORMATION, DEBILITY, AND DISEASES

OF THOSE

ORGANS PECULIAR IN EACH SEX:

INCLUDING

SPERMATORRHŒA, OR SEMINAL WEAKNESS:

DRAWN UP WITH A VIEW TO

THE TREATMENT AND CURE OF IMPOTENCE,

AND THE MORE EFFECTUAL

REMOVAL OF THE VENEREAL DISEASE.

THE RESULT OF THIRTY YEARS' EXPERIENCE,
WITH THE LATEST DISCOVERIES.

Illustrated by

NUMEROUS ANATOMICAL PLATES.

BY

WILLIAM WATSON, M. D.,

MEMBER OF THE ROYAL COLLEGE OF SURGEONS, ENGLAND, &c.

TENTH EDITION.

" Accepit nova fama fidem, populosque per omnes
Prodiit haud fallax medicamen."—FRACASTORIUS.

NEW YORK:

[Entered as the law directs.]

PREFACE TO THE ORIGINAL EDITION.

Among the great number of patients I have treated, many, through a humane desire to extend to others the benefit of that advice they had profited by, have importuned me to put into writing and publish a systematic work on the subject. To these some of the Profession have united their requests; suggesting as another motive, the number of books flying all abroad to misinform the public, written by persons without any pretentions to a regular education, and very imperfectly acquainted with the matter itself.

I will not conceal from the reader, that what especially had thus been urged by medical men, my friends, had long with me been the chief reason of silence; lest, in entering the same field with certain persons, I should be put in the same catagory with them. However, my reputation has now been too long established to be affected in that way; and I am at length content to lay before the world those discoveries by which so many persons have profited, and to whom it will be an additional gratification that I have heard their own request.

There is another scruple which I must declare has had less weight with me; it is the nature of the subject itself, thought by many of a kind that the public should not be made familiar with. What! are disease and vice of that nature, that we should rather allow them a silent, putrid course through society, than pollute the air, or hurt a fastidious public, by exposing and correcting them? This notion, certainly, has too much prevailed; for even the philosophic Hunter, giving way to "an unfeeling ribaldry," as Adams explains, suppressed some passages in his second edition, though "dictated by the purest benevolence." Mr. Curling, also, found himself obliged to append a note reflecting on Lallemand, whom he had followed; making use of his information, and then censuring his mode of communicating it. What the result is, let the same author express: "Taking into account," says he, "the repugnance felt to such inquiries, it is scarcely surprising that the subject has been but imperfectly investigated, and rarely treated of by the pathologist and practitioner."

It is evident, that either certain medical subjects are not to be alluded to at all, or they must be investigated with a philosophic scrutiny; and if the Profession is denied the liberty of speech, it inevitably comes that the empiric steps in, and does all the collateral evil, real or supposed, of publicity, without the advantage to counteract it of correct knowledge. The utmost decency of expression is certainly to be observed; yet what language can exceed certain practices it describes? or in what colours can we paint disease more hideous than its own deformity? The end, which is the cure, is a sufficient justification of the means we avail ourselves of to attain it :—

Per talia morbus
Tollitur, et nihil esse potest obscoenius ipso.

The present volume, indeed, is intended for secret perusal ; the subject of its very nature is private, as is the practice it treats of ; and since shame became an effect of the Fall, all people since indulge without a witness the most universal of all desires. When, therefore, I speak of publishing my researches, it is understood to be to every individual in the nature of a confidential communication, and sealed up, like a letter, for his use exclusively. Any other mode of address would shock the natural reserve of patients, who have a great apprehension of exposure. Nothing, to be sure, is more to be regretted than this sentiment, when it carries the sufferer so far as not to seek proper advice ; who would rather, like the Spartan youth, permit his bowels to be eaten into, than acknowledge the offence he is guilty of. Such subjects are not for every ear ; it is only the professional man who should be confided in ; and that professional man only, who makes a particular class of complaints his exclusive practice.

However, there are persons so situated, or whose fastidiousness is so great, that they will not address personally even the physician on the subject. Such must explain themselves in writing ; and for that purpose follow the Instructions to Patients, as given at the end of the volume. Among those who consult me, a great number meditate marriage ; some of whom prefer correspondence to an interview. One visit is always desirable, where practicable but it certainly may be dispensed with.

The plan of this work, divided into three parts, and treating of the formation, debility, and diseases of the genitals, the reader will find explained in the Introduction ; there is nothing more to add, and I will leave my little work entirely in his hands. I trust he will see in it a converse to the ancient Greek Epigram— "a great volume, a great evil ;" and that in the following pages he will find an inestimable treasure. Patients have often said to me, "I wish I had studied medicine ;" but here is all the result of the most extensive investigation fully unfolded, and a great deal more than most doctors are acquainted with. The invalid will find here the nature of his most indistinct sensations explained, the disorders, corporeal and mental, which they indicate, and the mode of curing them ; he will be taught to recognise the earliest signs by which disease or debility yet only threatens, and to know all those resources of skill by which the oldest symptoms may be removed ; he will find the opinions of the most eminent physicians, compared with, sometimes corrected by, my own ; and nothing, in fact, omitted that can conduce, whether to his health and longevity, to his vigour of body, or to his elasticity of mind.

TABLE OF CONTENTS.

GONORRHŒA.

CHAPTER I.

CHAPTER II.

SYPHILIS.

CHAPTER I.

SECTION I.
FIRST CLASS OF SYPHILITIC DISEASES.

SECTION II.
SECOND CLASS OF SYPHILITIC DISEASES.

SECTION III.
THIRD CLASS OF SYPHILITIC DISEASES.

¡SECTION IV.
FOURTH CLASS.

CHAPTER II.

CHAPTER III.

WOOD-CUTS.

INTRODUCTION.

The present Work I have divided into three parts : the *first* on the Anatomy of the Sexual Organs ; the *second* on Spermatorrhœa and Seminal Weakness, and the *third* on the Venereal Disease.

As I have studied brevity throughout, the Anatomical part will be found contained in a few pages, but yet I trust made sufficiently clear, with aid of the plates, which are of a superior description.

I have rested longest on the part treating of Spermatorrhœa, and Sexual Weakness, with Masturbation, as the most important ; and have compressed into the remaining pages the Venereal Disease, including its two chief divisions of Syphilis and Gonorrhœa.

Each part, however, will be found sufficiently full for the purpose of informing the Reader, who, I trust, will here derive all the knowledge he desires on these several subjects.

As a great many books are, now-a-days, mere compilations, that is, books written out of other books, permit me to say that the following is drawn entirely from actual observation, made during a course of thirty years and more, in which I have practised. Where I have found it necessary to state other men's opinions, I have drawn them direct from their works, having read them all through, and selected what was most to the purpose.

I mention this more particularly, because the earlier editions of this book, published many years ago, have been largely pirated from, quotations and all, by certain unprincipled quacks, forced to theft from the extreme poverty of their own knowledge. It is hardly necessary for me to add, that I have received a regular medical education, having served my time to a professional man, and taken my diploma for surgery more than thirty years since.

During the greater part of that time I have studied almost exclusively, and treated the complaints described in this book, of which I have cured many thousand cases. If, therefore, the Reader is so unfortunate as to require the aid of a Surgeon, he certainly cannot easily find one more experienced than the Author, or one in whom he can more safely rely. I am in constant attendance at my residence, and would direct such as wish to consult me to the "Instructions to Patients" at the end of the volume.

PART THE FIRST.

ANATOMY OF THE ORGANS OF GENERATION.

THAT part of the skeleton to which the organs of generation, or the chief of them, are attached in the male, and
which enclose them, at least for the most part, in the
female, is called the *pelvis*, from a fancied resemblance to

Fig. 1.*

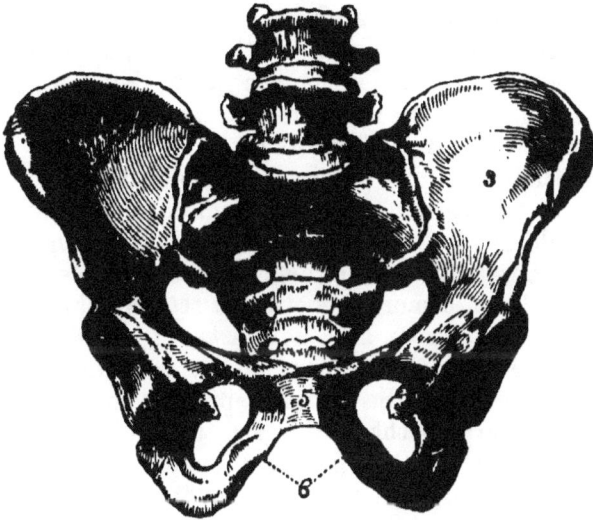

a basin ; although it has more the shape of a low chair or
seat, without the bottom : this, in fact, is the office it fills,
supporting the bowels, and the child in pregnancy ; for
which purpose it is larger in the female. The back part
is called the *sacrum,* which is but a continuation of the
spine, it is perforated to give passage for nerves from the
spinal marrow ; it ends below in a point, which projects
forwards. The plate of bone on each side, raised and
hollowed somewhat like a rustic seat, as I said, is called the

---

* The Pelvis. 1, the Spine ; 2, the Sacrum ; 3, 3, the Ilium ;
4, 4, the Pubis ; 5, the Symphisis ; 6, the Arch of the Pubis ;
7, 7, the Ischium.

ilium: this is the hip.   The front is the *pubis*, where the two bones are firmly united in the *symphisis.* 'Beneath this is the *arch of the pubis*, like the window in a Gothic building, except that it is less pointed in the female, to give passage to the child.   Lowest of all is that projection or tuberosity which supports the body when we sit ; it is named the *ischium*, and is a rough and stout piece of bone. The round, large opening observed on each side is simply a space covered with ligament and flesh, to save so much bone, of which Nature is very economical : beyond is the hollow to receive the thigh bone.

This cavity, as we style it, of the pelvis, appears very much open, and incomplete as seen in the skeleton ; but is very differently protected and enclosed before the ligaments, muscles, and membranes are removed.   The lower part especially of the pelvis is completed by a broad membrane, expanding from the ischium to the sacrum. The symphisis also is held by strong interlacing fibres. And the entire hollow of this arm-chair of bone, as I call it, is lined, and made more smooth and regular throughout.

I will not describe all the parts which the pelvis surrounds or gives attachment to ; a consideration too extensive and not in the design I propose myself.   I intend only the organs that propagate the kind ; every part of which I will now explain and make clear, with the smallest share of attention in the reader.

ORGANS OF GENERATION IN THE MALE.

These are the Testicles and Penis, with the Uretha, externally ; the Vesiculæ Seminales, or Seminal Sacs, and Prostate Gland, within.

The testes are suspended from the front of the pubis, in a loose bag, the outer layer of which is called the *scrotum.*   Next to this, like the lining of a purse, is the *dartos*, another sort of skin, with a greater ability of contraction, to sustain the testicle.   This double bag is enabled to contract very much and gather into a smaller compass, particularly in the robust, an evidence that may be taken of bodily strength, for in the feeble and old, it is relaxed and flabby.   In the dying, it is quite loosened, and one of the last symptoms before dissolution.   However, cold will make it purse up in all people, and excessive heat relax it in the most vigorous. Besides, it is under the influence of the mind when thoughts of the other sex prevail in it, and corrugates itself especially at this time.

The testicle, thus surrounded, a part in which Nature has shown an excessive care, has still a more immediate support, internal to these. This is a little muscle, or band of red contractile fibres, which pass in hooks round* the testis, and lift it up strongly, more especially in the sexual congress, when it is held quite near the body. Still, that it may move without the slightest interruption, this delicate body is farther placed in a smooth shining covering, infinitely finer than any cambric, well moistened with an exhalation from its surface. It must not escape notice that the left testicle is hung lower than the right, lest in closing the thighs they should receive injury.

Fig. 2. †

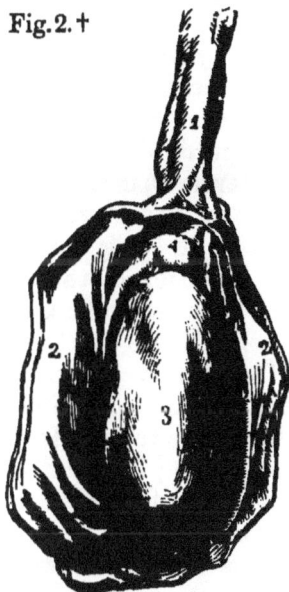

With so much protection external to it, we are prepared to meet the most elaborate complication within. That this delicate part might preserve a suitable form, it is bound in a strong elastic capsule; which, in addition, sends in several bands in a manner to divide the testicle into so many different compartments, and at the same time to brace it against compression. There is a principal partition to which all these minor ones are conjoined, which is united with what may be called the back part of the testicle.

It is within these spaces that the proper structure of the gland may be noticed, and what makes the bulk of it. This is composed of a countless number of minute tubes,

* See Fig 10, numbers 4, 4, 4, page 20.
† Testis, with part of Spermatic Cord ; the coverings of the Testicle are laid open. 1 Spermatic Cord ; 2, 2 Tunica Vaginalis ; 3 the Testis ; 4 Epididymis.

each about $\frac{1}{150}$ of an inch in diameter, and which may be drawn out as much as two or three feet, although no doubt they are longer. Each of these little tubes is coiled so closely upon itself that it appears knotted ; and, also, many of them join into a bundle, of which bundles of tubes there are not less than three or four hundred.

Fig. 3.*

These tubes may be said to commence at the round of the testicle, from which they approach to the back part, uniting with each other, in a way that two or more go to form one, which again conjoin to make a less number ; until, successively coalescing, these numerous tubes are at length brought down to about twenty, only larger than the primitive ones. These twenty traverse for a short distance in straight lines, reaching near the back of the testis, where they reunite, and reduce to eight or ten. These eight or ten, finally, twisted in the same extraordinary manner, emerge from the testicle ; they now are increased in size, and form into a single tube. This single tube does not cease to have the same tortuous disposition ; it is laid along the back of the testicle in a close coil, to which anatomists give the name of *Epididymis.*

It is from the lower part of this epididymis that the tube ascends, without further alteration, until it enters the body *above* the pubic bone, through an opening made for it in common with the artery, vein, and nerve which supply the testicle, called conjointly the *Spermatic Cord.*

This lengthened description the reader will more fully comprehend by the image of a large river (to illustrate small things by great), commencing from innumerable latent springs in the elevated country, which supply separate rills, one by one uniting to form a less number as they

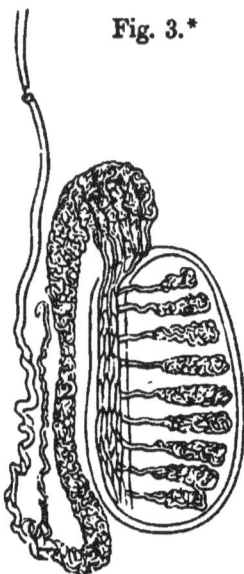

* Anatomy of the Testis. 1, 1, 1 the Tubuli Seminiferi ; 2, 2, 2 the Epididymis ; 3 the Vas Deferens.

descend, and returning often upon themselves, like the Meander; until, making a few larger rivers, these, at length, pour all their floods into a single one.   And, indeed, literally in this way it is, that the seminal fluid, elaborated from the blood, is collected, and poured along in a single stream, in the end, to be conducted internal to the body, and then sent down to the root of the penis, in a way that I shall now describe.

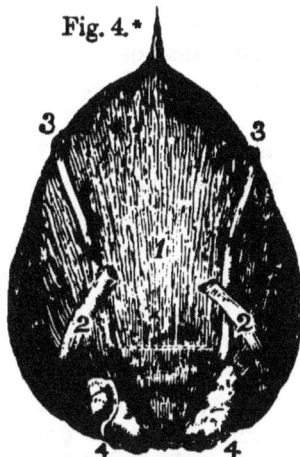

Fig. 4. *

This tube I have mentioned, having left the testicle, assuming the name of *vas deferens*, and passing, as I said, into the body, not beneath, but *above*, the pubis, turns downwards along the side of the bladder until it gains near the neck of it, where it meets what is called the *prostate gland*, through which it passes, and opens into the urethra, or that tube which gives passage from the body to both semen and urine.

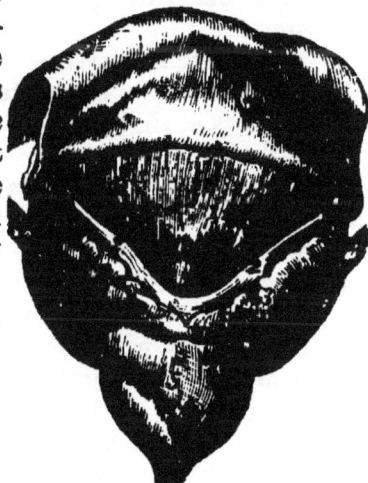

Fig. 5. †

This prostate is a firm, little yielding substance, somewhat the shape of a chestnut, surrounding the neck of the bladder, and conveying through it the urethra entering the penis. It is hollowed in small cavities, with openings, through which a brownish fluid is pressed

* The posterior aspect of the Male Bladder. 1 the body of the Bladder; 2, 2 the Ureters; 3, 3 the Vasa Deferentia; 4, 4 the Vesiculæ Seminales.

† The under surface of the Base of the Bladder, with Vesiculæ, and Prostate. 1 Bladder; 2, 2 Vasa Deferentia; 3, 3 Vesiculæ Seminales; 4, 4, Ureters; 5 Prostate Gland; 6 Urethra.

out before the urine, but especially before the semen, to lubricate the way, and allow it to pass more easily : a sort of anti-attrition, to use a term in mechanics. It is very small before puberty, but apt to enlarge in old age, and bring on a very troublesome malady.

On the sides of the prostate lie the *Vesiculæ Seminales*, in shape oval, and about two inches long. These vesiculæ are each like a testicle of a less complicated structure ; being simply a tube much convoluted, only communicating frequently with itself, by small openings, where it is doubled. This tube, separating from the rest of the vesicula, opens into the vas deferens, or tube from the testicle, a little before it joins the urethra, as just described, through the prostate.

Fig. 6.*

The vas deferens increases in capacity as it passes the vesicula, and becomes sacculated, as is the vesicula itself ; but it narrows again before it joins to form a common tube with the duct of the vesicula. This common tube is called the *ejaculatory duct*, or *common seminal duct ;* it is less than one inch in length, and opens in the floor of the urethra, close to that of the opposite side ; or, rather, opens into a depression, or small sack, of the prostate, about a quarter of an inch in length.

If there be any part meriting a more scupulous investigation, and of greater importance, it is this narrow space ; for the ejaculatory ducts opening into it, form a sort of flood-gate to the semen, and allow it to escape passively, which is the diseased state, or hold it in, and reserve it, for occasion, then to escape in a collected body. There is a slight fold of membrane, called *caput gallinaginis*, which separates these openings, and also overlays them. The sensibility around these openings is at once great and peculiar ; without it, the sexual sensations could not pro-

* Vesicula Seminalis, Vas Deferens, and Ejaculatory Duct, laid open. 1 Vas Deferens ; 2 Vesiculæ ; 3 Ejaculatory Duct.

ceed; for it is the centre of a round of actions, in place as well as in function, and everything turns upon it. Aroused too often, it is dissipated altogether; or is strained to so morbid a degree, that it fails in a nervous irritation. Of exquisite endowments, man provokes this precious part with the most libertine disregard; for which his only excuse is, a want of that knowledge on the subject which I am now endeavouring to supply him with.

The urethra is the next part the reader will expect the description of; in which I will gratify him presently, when I have explained, in order, some other parts of the penis.

Fig. 7.*

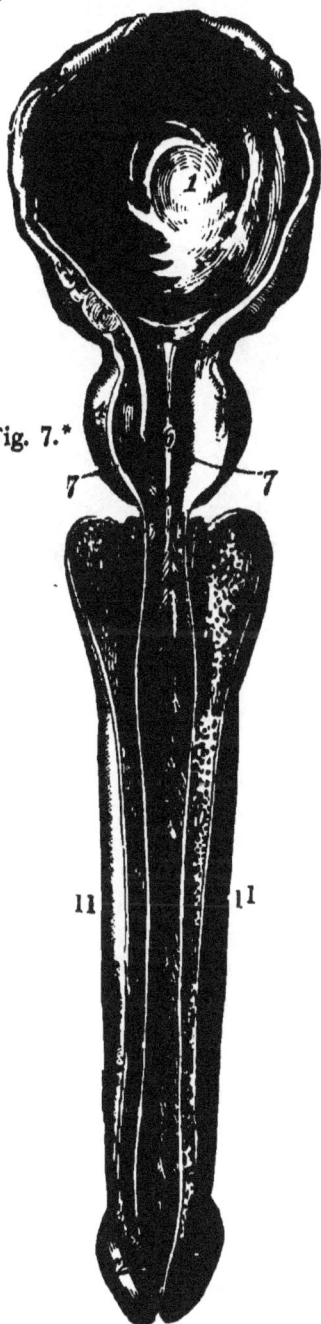

* The Urinary Bladder, laid open from the front; the Urethra is also exhibited through its entire course. 1 The Bladder; 2, 2 Openings of the Ureters; 3, 3 Neck of Bladder; 4, 4 Prostate Gland; 5, 5 Prostatic part of Urethra; 6 Verumontanum, or Caput Gallinaginis; 7, 7 Openings of the Ejaculatory Ducts; 8, 8 Cowper's Glands; 9, 9 the Bulb; 10, 10 Corpus Spongiosum; 11, 11 Corpus Cavernosum; 12, 12 Glans Penis; 13 the dilatation called the Fossa Navicularis; 14, 14, the Urethra.

The *penis* is fastened firmly to the bone, by two *crura*, or limbs, from the lower part, or tuberosity, of the ischium, up as far as the top of the arch, or symphisis of the pubis, where they unite and form the body of the organ, which, however, continues to be marked vertically into lateral parts. These crura, thus united, compose a strong tendinous tube which is subdivided into a great many cells.

Into these cells it is that the blood enters during certain impressions from the mind, or on mechanical irritation, producing erection.

Fig. 8.*

These cells are supported by fibrous bands, at intervals, across the organ, which I know are necessary to maintain sufficient firmness in it, from a case in which I was consulted. The penis, which otherwise was sufficiently strong in this man, was so weak towards the middle as to bend suddenly like a joint. It was readily doubled in this way during erection, but instantly rose again when let go. He complained that often it lost its place in the vagina. The want of firmness at a particular part only, can not be explained otherwise than by supposing the transverse bands broken or deficient thereabouts.

Underneath, all along, where these lateral tubes of the penis are joined, is left a grooved space, into which is laid the *urethra.* It is imbedded in a fine erectile tissue, of a similar nature to that of the penis, distended by the same cause, and simultaneously with it. The posterior part of this tube, behind the scrotum, is overlaid by two little muscles on each side, which, contracting, expel the semen, and the last drops of urine : they are called *Compressor Urethræ,* and *Ejaculator Seminis.* It is upon the energy of

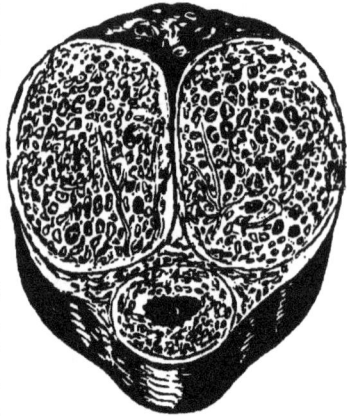

* A Transverse Section of the Penis, made at about its middle. 1 The Fibrous Sheath of Corpora Cavernosa ; 2 the Septum ; 3, 3 Cells of Corpora Cavernosa; 4, 4 Corpus Spongiosum; 5 the Urethra; 6 Vessels and Nerves of the Dorsum of the Penis.

these muscles that the success of *coitus* so much depends.

In front of the penis is the *Glans*, which is but a continuation and expansion of the spongy structure in which the urethra lies, and which is thereby distended along with

Fig. 9.*

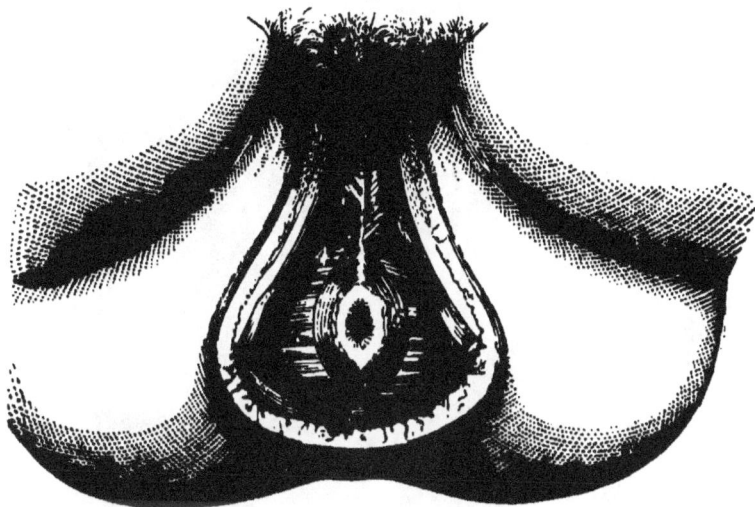

it. This glans is closely united with the body of the penis ; but the blood does not pass freely from one to the other, which, consequently, may be distended separately.

The root of the glans rises in a ridge above the surface of the penis, leaving a groove, over which a fold of the skin is continued, long enough to come as far as the opening of the urethra, and sufficiently loose to be drawn completely backwards. This is the *Prepuce* ; and is held underneath by a delicate bridle called the *Frenum*.

The testicles, in some rare instances, seem wanting when not really so. This may be easily explained. In the child, during most of the time before birth, the testicles are laid a little below the kidneys ; from which they descend, and come into the scrotum. This descent, however, does not always occur ; sometimes one, sometimes

* The Perineum. 1 The Acceleratores Urinæ Muscles ; 2 the Erector Penis Muscle ; 3 Corpus Spongiosum ; 4 the Transversus Perinei Muscle ; 5 Sphincter Ani Muscle ; 6 Levator Ani Muscle.

both, remain in their original place, and are never seen. A controversy has arisen, but it seems to be decided that position does not alter the power of the part.

Among so many persons as have consulted me, I have seen numerous irregularities in the formation of the penis ; of these, the most common is the bending of it when excited, in some, to one side, in some, to the other ; likewise, upwards or downwards. The prepuce, in many, is drawn so close, like a purse by the string, over the front of the glans, that it is never seen. The most usual malformation is in the opening, set too far underneath, or with the appearance of a cleft.

Fig. 10.*

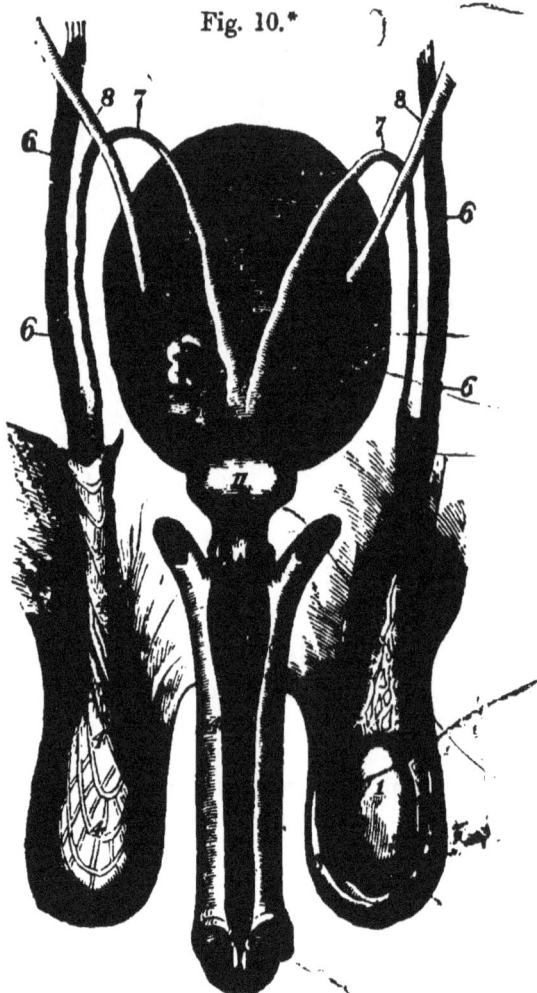

* General View of the Male Organs of Generation. 1 The Testis; 2 the Epididymis; 3, 3 Coverings of Testis; 4, 4, 4 Cremaster Muscle; 5 Spermatic Cord; 6, 6 Artery, Nerve, and Vein ; 7, 7 Vasa Deferentia ; 8, 8 Ureters; 9 Bladder ; 10, 10 Vesiculæ Seminales ; 11 Prostate Gland ; 12, 12 Cowper's Glands ; 13, 13 Crura Penis ; 14 Corpus Cavernosum Penis ; 15 Corpus Spongiosum Urethræ ; 16, 16 Glans Penis.

ORGANS OF GENERATION IN THE FEMALE.

The *Uterus* is of a triangular, or pyriform, shape; indeed, not only the shape, but about the size of a large pear,

Fig. 11.*

being about three inches long, and two broad at the fundus; while at the neck, which is the anterior part, it is narrow enough to project into the vagina, which a short way embraces it. The mouth of the uterus, or *os tincæ*, appears as a narrow slit in this projection. The passage through the neck is correspondingly narrow, and the cavity of the womb itself narrower than might be imagined, occasioned by the thickness of its sides, which is about an inch.

The direction of the uterus is oblique, upwards and backwards; from the corners of which ascends, on each side, the *fallopian tube, or oviduct;* about four or five inches in length, they turn and make an arch downwards. They are about the thickness of a goose-quill, but the passage through them is so small as scarce to admit a bristle, except at the end farthest from the uterus, where it is expanded, as it were, like a diminutive trumpet. This expansion is irregular and fringed, receiving the name of *corpus fimbriatum :* it overhangs the ovary, and is connected with it by a few short fibres.

* The Uterus, with its Appendages, viewed on their anterior aspect. 1 The Body of the Uterus ; 2 its Fundus ; 3 its Cervix ; 4 the Os Uteri ; 5 the Vagina, with the Transverse Rugæ ; 6 Ligament ; 7 Fallopian Tubes ; 8 Fimbriated Extremity of Fallopian Tubes ; 9 the Ovary.

The *ovaries* are two small oval bodies, divided into cells. Placed in these cells are certain small vesicles, from six to twelve in number, about the size of a small pea, containing a yellowish fluid. These are not the ova, but the sacs in which they are contained; for the ovum itself is not more than the $\frac{1}{160}$th of an inch in diameter, with still a little spot within it, supposed to be the actual germinal substance.

PART THE SECOND.

ON SEXUAL DEBILITY, SPERMATORRHŒA, OR SEMI-
NAL WEAKNESS ; TOGETHER WITH MASTURBA-
TION, AND THE BEST TREATMENT OF IMPOTENCE.

CHAPTER I.

NERVOUS STRUCTURE OF THE TESTIS ; WITH INFLUENCE OF THIS ORGAN ON THE DEVELOPMENT OF THE BODY, AND THE SANITY OF THE MIND.

NOTHING is better known to physiologists than the influence which many parts of the body have over the remainder. The reason for this sympathy is the universal
interlacement of the nerves, which originate and transmit
sensation mutually among all the organs. They are connected in the brain, as a centre, which itself is but a large
agglomeration of nervous fibres, interlaced, folded, and
packed away among the numerous recesses of the cranium.
It is here that the more various and subtile operations
and sympathies of the mind itself are put in motion,
through a machinery and relations never to be explored :

"Quod latet arcana, non enarrabile, fibra."

Beside the general distribution of nerves, there is a
special network more visible, which invests each of the
more important organs, and conducts its functions more
immediately. This we call a plexus or reticulation, in
which the threads traverse, unite, reunite, and intertwine,
in a manner altogether inextricable. But, of all the
organs, none is more obviously provided in this way than
those of generation, which form a most delicate and intri-
cate piece of anatomy. Has the reader seen the ball
of lace at the Great Exhibition, over which so many
thousand threads were drawn, and knotted into the web ?
This is but an insufficient illustration of the nerves as
they surround the testis, as also the ejaculatory ducts, and

glans penis. When describing these parts, I purposely omitted mention of the nerves, partly that I might not crowd the description, but chiefly that I might introduce the matter here, to account for the wonderful influence which these parts have over the rest of the economy.

As the organs peculiar in each sex are not developed until after a certain number of years, we have an obvious illustration of the effect they produce at maturity. Children are scarce to be distinguished among at first, and for many years go about together in unconscious innocence; of a sudden, however, the change comes; the organs exclusive in each are evolved, and then operate on the entire system; the beard and rough voice declare the man; the breasts the woman; but the eye speaks the most. The sexes recognise one another at a glance; which discloses, what cannot be concealed, the strange revolution in the mind itself, brought about by the same action of a distant part of the body over the whole.

That it is from this source so strange a series of alteration follows, we have proof sufficient. There is an obvious submission in the creature from which the testicles especially have been taken. How different is the docile gelding from the horse unbridled and entire, say, as in the description of Homer :

> The wanton courser thus, with reins unbound,
> Breaks from his stall, and beats the trembling ground;
> Pampered and proud, he seeks the wonted tides,
> And laves, in height of blood, his shining sides;
> His head now freed, he tosses to the skies,
> His mane dishevel'd o'er his shoulder flies;
> He snuffs the females in the distant plain,
> And springs, exulting, to his fields again.

Man is not less seriously affected by mutilation, seen principally in the " big, manly voice," which continues tenor through life, as in the Italian *castrati*. As the horse is without the crested neck and flowing mane, the man is without the beard, and, which is very remarkable, swells out, like a woman, over the hips, with a rounding generally of the outline.

With so great a power, therefore, over mind and body, we should expect to see it more remarkably when those parts are irritated, or put in action too frequently.

Some animals, indeed, in the lower scale, seem to have little other purpose in their being, than to do the deed of kind, as they perish a short time after, from exhaustion. A phenomenon analogous to this may be mentioned of some plants, which die as soon as they have flowered; but if the sexual parts are removed, they survive another year, and are prolonged to a biennial existence, by sparing them the weakness from fruition. I have known a single sexual effort induce epilepsy in those never attacked before. I remember an instance of this sort, in which I was consulted. It was a youth, nineteen years of age, who was admitted by a young woman several years older into her room two or three nights in succession. It was the first time, he assured me, that he had tasted of these pleasures. Not so, we may suspect, the other, who spurred the young fellow on to such a degree, that of a sudden he fell into convulsions, which so much terrified his paramour, that she screamed loud enough to alarm the family, and thereby disclosed the whole matter.

In addition to the immediate violence, there is a general wasting of the flesh; so that it has been estimated, that the loss of half an ounce of semen, is equal to a pound of blood: not that it is the waste of fluid only, but the exhaustion of vital energy. Those who have written on consumption take notice how many die of it a little after the time of puberty; which is easily accounted for by the overindulgence of a new-felt passion. The chief strength of the body, it has always been supposed, is in the back; whence the expression, "the strength of his loins," used in the primitive ages, and even that "out of the strength of his loins" a man's children proceeded. From observing, therefore, the lassitude in this situation, from such causes as I am speaking of, the early anatomists thought the seed a solution from the spinal marrow. Others, again, whose attention was drawn to the effects in the head, supposed the brain secreted this fluid, through some imagined channels.

Not that the body in general is that which always suffers more evidently. The parts themselves are more or less deficient of energy; according to the excess of irritation, or power in a strong constitution to resist it. The impulse returns at longer intervals; sensation is less vivid; and the semen propelled too hastily. The progress is always slow and insidious; but this last mentioned symptom is the earliest; nearly all those patients who consult me mentioning it more emphatically. What heightens the disappointment is that it is mutual with another, whom no anxiety to gratify can delay the impulse for an instant. How many have told me that this was the accumulation of distress! when the baffled partner to enjoyment remains unsuspecting, perhaps only with that instinct of expectation taught unerring by nature. The languor natural to the occasion is depressed into dejection; and the physical incapacity increased by a mind conscious that the attempt was unsuccessful. By degrees, the testicles are flaccid, or wasted, with a low animal heat; and erection of the penis less energetic, and less lasting. This is one of the chief symptoms the patient remarks; who may observe also that the glans is not distended, although the body of the organ is sufficiently so. An erection of this imperfect kind may be observed in children; and it may continue long after the real venereal priapism has ceased to return. The pressure of water on the neck of the bladder will cause it; which accounts for what patients have often mentioned to me, that they have erections only at getting up in the morning.

In this state the patient may continue for years, unassured, and with little else than the cheat of pleasure. But, as hope deferred maketh the heart sick, despondency begins, and it is in the mind we are to look for the principal evidence of this malady. "There is perhaps no act," says Hunter, with as much knowledge of morals as he undoubtedly had of medicine, "in which a man feels himself more interested, or is more anxious to perform well, his pride being in some degree engaged;" and again, "nothing hurts the mind of a man as much as the idea of inability to perform well the duty of the sex." Yet it is not chiefly from unpleasing reflections that these peculiar effects take their rise; it is rather from that species of influence I have described which a remote part of the body may have over another, and which the sexual parts have so remarkably over the brain. Nothing can be stranger than this, but nothing more true, or less contested in physiology.

It would appear that the healthy secretion of semen is essential to sustain a kind of elasticity in the mind; which, otherwise, droops, and sinks in a settled melancholy. From this animal source comes that excess of spirits which animates in the career of life, and prompts us to look still onwards, with a brighter prospect. In proportion, therefore, as this fountain is dried, anticipation fails also; the allurement of the future is lost; and all the fair horizon of society, dark and clouded. It is as if all the colours were taken out of nature; and the eye discovered that all those images which beautify existence were not in the objects, but in itself. Every purpose of life appears idle, ambition, gain, and honour; and melancholy haunts in the soul.

After the reflections here given, the want of support in the mind itself, and that void which we have seen spread before it, from the morbid effects of irritation in the spermatic organs, it will not surprise the reader that the intellect may fail in other ways, more or less intense and remarkable. With numbers there is an apprehension altogether vague and undefined; or apt to associate itself with any occasional grievance or terror : at one time, with the uncertainties of trade; at another, with changes in public opinion; now, it unites with the prevailing epidemic, and one thinks himself immediately liable to the cholera, whose real malady is weakness of nerves, from another cause.

The most common form, indeed, is what is called nervousness; excess of spirits and gloom alternately; a certain agitation from slight causes, intolerance of noise, desire of change ; together with petulence, distrust, and discontent. Various minds are variously affected ; some become more solitary and morose; while others have recourse to confidence, which they often bestow injudiciously. For the most part, however, there is a want of animal courage; and although the quickest perception of an affront, a pusillanimity to resent it. Application is continued with difficulty, and the memory is weaker. Association becomes irregular ; one thought suggests another without any obvious relation, which again darts in a direction entirely unexpected. At length, some one idea settles in the moveable foundation, and remains the fixed point of error. After a hundred hallucinations have chased one another, like shadows, perhaps one's nearest friend is distrusted, and then every thing is inverted around him. The worst is, that while perception perhaps fails, and the memory grows weaker, reason is morbidly acute and pursues the concatenation of thought with a strange intensity. It will vindicate its conclusions with the greatest address; and escape detection in a fallacy, with a cunning peculiar to the insane.

Until of late years this source of lunacy had been completely overlooked, from I know not what strange fastidiousness in medical men to investigate it. For this reason, the empiric knew more than the regular practitioner on the subject, and, although without the lights of science, had those of observation to direct him. At length, from the growing evil, in a licentious age, it was forced upon the attention; and especially in America has been considered with statistical precision. In the commonwealth of Massachussets they have ascertained, that while aberration of mind is more frequent than in any European country, it is chiefly owing to seminal irritation, produced by artifical causes. (*See* Report on the subject of Idiocy, Lunacy, &c., to the Senate, by Dr. G. S. Home). It is remarkable that an eminent French writer who has treated of this subject, had it brought indirectly to his notice : for having issued a work on cerebral affections, a great number of patients were sent to him, to be treated on that supposition; whom, upon investigation, he found subject especially to diurnal pollutions. And, not to go beyond my own experience, I have over and over again seen the mind wavering, and ready to fall in those who consulted me on this complaint; which, there could be no question, was the cause of the other evil, as they disappeared consecutively, the physical capacity first restored, and then the mind gaining all its elasticity.

By stages such as those I have described may the human mind be perverted; and in such a manner may the body be altered, as explained previously. I have drawn but the outline; though I might easily fill up the picture, and present a more mortifying spectacle of humanity in the ruin. The reader may rely upon it that I have copied from nature; that is to say, that I have taken the description from the many histories patients have given me, in the different degrees, shapes, and complications, of their cases. Such as it appears, the philosopher's definition of happiness is inverted, *mens sana in corpore sano;* for here the nervous irritation of the frame extends itself to the understanding, and overthrows it also.

But it is now time that I should consider this most serious subject more minutely; that I should explore the immediate nature of this irritation whence it comes; the different causes that produce it; the effects, as they follow; and thence, ultimately, the cure.

HISTORICAL NOTICE.

I WILL only ask the reader's indulgence through one or two pages: who perhaps will not be indisposed to trace the historical account of this malady, so little known, until very lately, to the moderns, and so well understood by the ancients, those accurate observers of nature. Hippocrates, whom we call the Father of Medicine, has taken notice of it; who, although mistaking its real nature, has very accurately described some of the symptoms. He calls it " dorsal consumption, which," says he, " arises from the spinal marrow; it affects chiefly the newly-married, and libertines. They are free from fever, and eat well; but they lose flesh. When they pass water, or go to stool, they pass much liquid semen; and want the power to impregnate. They have emissions in their dreams, whether sleeping with a female or not. Walking, or running, especially if ascending, they experience suffocation, lassitude, weight of head, and noise in the ears."

The next classic author is Celsus, the Latin Hippocrates, who has this passage: " There is a disorder incident to the private parts, a waste of semen, without venery, or nocturnal dreams, which, after a time, induces leanness and consumption." He seems to have taken up the description where the Greek left it, and carried it into the more advanced stage, which is what we call now-a-days " passive spermatorrhœa."

Aretæus dwelt more on the general symptoms: " The flow of semen gives an appearance senile, slothful, enfeebled, timid, dull, silent, imbecile, wrinkled, inactive, pale, whitish, effeminate, chilly; with aversion to food, heaviness in the limbs, torpidity, impotence, and languor in all things."

Fabricius of Aquapendente notices the fluidity of the semen, which he supposed found its way on that account more readily out of the vessels.

Aetius: "Disorder of stomach ensues, with weakness of body, pallor, leanness, shrivelled skin, and hollow eyes."

I have already noticed, that some writers had supposed the seed a solution of the spinal marrow: this will make intelligible the description of Tulpius, who speaks of wasting of the spinal marrow. "The body and mind," he adds, "equally languid, and the man perishes miserably." This wasting of the medulla mentioned by ancient authors is, in effect, the same as what we call softening of the brain and spinal cord; a well-known result of seminal weakness; although it was a mistake to conclude the semen a sort of discharge from the nervous substance. These authors give the facts accurately enough; the error was only in the hypothesis.

The famous Ambrose Paré describes the flow of semen, which, says he, is always whitish; it is the cause, he adds, of marasmus, excessive prostration, and a sickly hue over the body : the discharge is without titillation or pleasure.

Hoffman alludes to sleep which does not restore, with broken dreams. Nor did the illustrious Haller overlook this malady; mentioned also by so many other writers, superfluous to quote.

One, however, I will select, Sauvages, who, in his Nosologica Methodica, speaks of the semen "escaping habitually, as a momentous evil, obstructing generation. This happens," he continues, "from the excretory mouths of the visculæ being relaxed or eroded ; caused by repeated abuse of these organs, as happens to filthy masturbators, most of whom are unable to retain their seed afterwards, and pour it forth, without sensation, during evacuation of the bowels. This happens also from repeated virulent gonorrhœas, in other respects cured, and all other effects removed. From such frequent and untimely emission of semen many diseases arise, but chiefly hypochondria, asthma, tabes dorsalis, sterility, anorexia, sleeplessness, epilepsy, loss of memory, amaurosis, flatulency, pallor," &c. Sauvages alludes presently to *Dyspermatismus Serosus*, which he comments on "as an ejaculation of watery semen, unfit for generation, and the most frequent cause of sterility in man. Semen of a very watery kind is poured out during coitus, and in no small quantity, but with very imperfect and temporary erection."

At length Tissot, whose great reputation spread over Europe, and, as Gibbon mentions, peopled Lausanne, wrote a work exclusively on this subject. He describes minutely what had escaped all the Ancients, the deplorable effects of masturbation; whether this vice was unknown to Greeks and Romans, or only introduced with the reserve and secresy of modern manners. For it is obvious that as society now is, and female chastity guarded so preciously; and, also, as the venereal disease, unknown to antiquity, at least in those terrible shapes which are familiar to us, so much deters against the courtezan; not to mention the passion for boys, so universal in the pagan ages; a gratification is sought, or shall I call it, outlet to a passion, which while it cannot be said to have introduced a new malady, has aggravated, and made more general, all the symptoms of it. Tissot exposed and explored this new vice in morals, and cause of disease in pathology, who, notwithstanding his numerous errors, has put an everlasting obligation on mankind.

"Masturbation," according to Tissot, "is the cause of a number and variety of evils: Total derangement of the stomach, shewn in some by loss or irregularity of appetite; in others, by acute pain, especially during digestion; by habitual vomiting, resisting all remedies so long as the bad habit is continued: Weakening of the respiratory organs, whence frequently result dry coughs, almost always colds, weakness of voice, and sense of suffocation on slight exertion: General relaxation of the nervous system: Extreme debility of the organs of generation; the greater number complaining of imperfect erections, the semen escaping as erection commences, or just as it is complete; others have no longer desire, but fall into complete impotence. Nocturnal pollutions are a terrible scourge to them; and often overwhelm those whose organs are absolutely senseless when they are awake. When these patients have had nocturnal pollutions, they find themselves the following day in a state of depression, of discouragement, wearinesss, misery, lassitude; pain is felt in the loins, stomach, head, and eyes; all which renders them truly pitiable, while they are changed to that degree as to be scarcely recognised."

From the days of Tissot to our own, here and there some eminent authority adverts to this species of malady : Frank, De curandis hominum morbis; Wickman, De pollutione diurnâ; Sainte Marie, who translated Wickman; Serrurier, and, to omit others, Deslandes.

Lallemand was the last to arise; whose work Des Pertes Seminales Involontaires has once more given prominence to this subject, and forced it upon the notice of the age. For this alone a great debt is due to this author; although his work is but a crude indigested mass, in which, while he affects a semblance of method, there is nothing but repetition and irregularity. His pages are over-crowded with cases, which present the same unvarying symptoms, adding much to the reader's fatigue, but nothing to his knowledge.

CHAPTER II.

ON MASTURBATION, AND SOME OTHER CAUSES OF SEXUAL INCAPACITY.

It may naturally surprise the reader unacquainted with the nature of morbid irritation, that a part so small should have so wide and powerful an influence over the remainder of the economy. It must be understood, however, that sensibility is not to be measured by mechanical rules; and

Fig. 12*

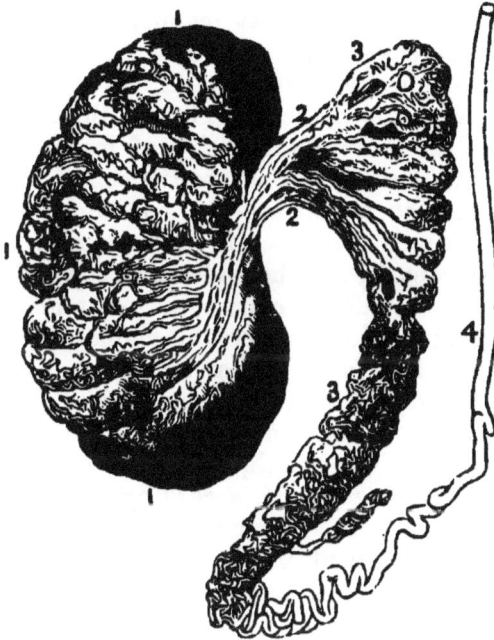

that a minute organ with high vital endowments, or even part of it, excited beyond proportion, may exasperate the

---

* Glandular structure of Testis, displayed by mercurial injections. 1, 1, 1 Testis, as subdivided into lobes, each lobe being composed of convoluted tubuli closely packed; 2, 2 Vasa Deferentia; 3, 3 Epididymis; 4 Vas Deferens.

frame to a much greater degree than a larger one, but with a lower vitality.

However, even in that view of it, the extent of surface enclosed within the testicle is a great deal more ample than could be imagined. Lauth makes the seminal tubes 840 in number; and calculates the entire length of all united as 1750 feet. I am inclined to go beyond this estimate; and considering the great number of the seminal tubes, the fineness of their coats, their extraordinary convolutions, and the studious endeavour of nature to coil, and pack them away (in that manner fully explained in a former chapter)—I believe if they were all opened out on a level, they would cover an acre. Let this not astonish the reader who knows the conjecture of Sir Isaac Newton, that, from the extreme porosity of matter, the whole earth would allow of pressure to the size of a cubic foot, before it came to actual solidity. And perhaps also the classic reader will remember, in illustration of this matter, what Pliny reports, that he had seen a copy of Homer in a nutshell: a volume in reality the most ample, as it is, also, the most precious property of mankind.

Along all this line of tubes the seminal fluid is constantly secreted; where it remains partially, and is partially absorbed, to be remixed among that blood from which it comes. In health, this fluid accumulates in a few days, or a few weeks, according to vigour; and is the cause not only of sensation immediately where it lies, but also of excitement in the peculiar structure of the penis, together with a heat of passion, which suggests very naturally its object.

Nature, therefore, wanting that only way at first designed, every artificial provocation should be avoided. The worst of it is, that in proportion as those obstacles to a commerce with that other sex which is formed for it accumulate, those circumstances that incite to it gather also; and the youth or man is constrained between opposite necessities. For instance, among all the artificial gaieties of a ball-room, the studied effect of dress, the generous heat of wine, the agitation of dancing, with the "lascivious pleasing" of music; what situation more likely to rouse every latent disposition, or less so, in the strictness of modern decorum, to give way to it?

The repetition of an exciting scene like this soon palls upon the adult, and has its own peculiar inconveniences. Not so the youth, who retires to his chamber from this tumult of all that is alluring, "dazzled and drunk with beauty;" he betakes himself to his bed, and there indulges his imagination to the utmost. He was taught perhaps earlier than this, perhaps yet at school and unsuspecting, a false art to pleasure, but a real one to destruction. This he has recourse to, and disburdens for a time an over-charged mind, and a heated organ: a mind and organ of reciprocal influence, and acting more violently in the concussion.

This unnatural trick once practised, is too apt to be repeated, like other vices ; perhaps it was not known to be a vice, or not suspected of the mischief it produces. However, at first used because the course of nature was not accessible, use becomes a second nature, and masters it; so that what was sought before as a substitute, is now preferred : so much ascendency can fancy aspire to. But not only this revolution to a depraved taste, there is also a physical change as decided ; the parts are changed in sensibility ; and having repeatedly received impressions of an artificial description, are less susceptible of those only they were at first formed for.

In this way, when masturbation has been practised for a number of times, the passages become incontinent of their fluids, and even to eject them in sleep: not through the superabundance of secretion which I have already spoken of as seeking an outlet ; but through mere weakness, and inability of retention. This emission succeeds as an effect from the other or voluntary pollution; usually with increasing prostration ; and often without erection of the organ. The mind and body sympathise mutually ; the one is relaxed, and the other impaired in all those varieties of symptoms I have previously noticed.

This waste of the seed which takes place usually at night, when the body is heated by much covering, occurs in some irritable habits in the day, from any mechanical irritation. I forbear to give the particulars of cases, which would swell these pages to as many volumes ; but I have had many patients who told me that the jolting of a carriage, the shaking of a horse, or the mere pressure of the clothes in walking, would provoke to ejaculation. The tremulous motion of a steamboat produces the same effect, and even the alternate rise and fall in a swing.

So long, however, as the fluid is thrown out in a jet, whether from provocation or spontaneously, the debility is so much the less, as it shews some power in the seminal tubes to retain the fluid, and some vigour in the muscular fibres to propel it. But in more advanced stages of the malady, emissions are less frequently seen ; the tubes eliminating and forming the semen, are not able to retain it ; and it flows away insensibly, without drawing the attention of the patient. Sometimes it will leak outwards, and stain slightly the linen ; or it may find its way backwards into the bladder, and mingle with the urine. The urethra is relaxed, and readily permits the fluid this direction. In such as have the secretion abundant, it will shew as a sediment in the urine; but these are not the most numerous, for it is more apt to be scanty and without consistency, so as to remain diffused through the urine, and invisible to the unaided vision. By the aid of a powerful microscope, however, and after some practice in the use of it, the spermatozoa may be readily detected, with their peculiar appearance and organization, as I shall describe in a following chapter.

In extreme cases, however, these animalculæ cannot be detected, which are confessedly the active principle of the semen. The testicles have no longer the power to form them ; and the fluid is thin and watery, leaving scarce any stain upon the linen, or trace in the urine. Little susceptibility to impressions remains in the penis ; but the mind may have a strong influence still ; and the presence of a female be fully recognised. The fluid now comes away with a cold depressing sensation ; and the lassitude of coition is felt, without the effort or the pleasure of it. Many, it is true, lose the thought with ability to venery ; spend their cheerless days and cheerless nights in apathetic despondency ; but with others, lascivious ideas still haunt the mind, the ghosts of former sensations; the slightest thought or passing glance, a picture, some passage in poetry or fiction, will exalt the mind, and bring all the uneasiness of uncomplying desire. A sexagenarian, who had led a very libertine life, once consulted me, whose very eye spoke lechery; but he had quite dissipated his virility: "Woman is the importunate subject of my thoughts," said he, " but it all rests here," pointing to his head, " it rests here."

That extreme irritability of the seminal tubes, with irritation of the fluid they elaborate, which I have been speaking of, has, with a circumscribed and unphysiological view of the subject, been confined to one short part alone of them ; or that which anatomists call the ejaculatory ducts, which are the tubes extending something less than an inch from the vesiculæ to the urethra. These parts, no doubt, are much concerned, and must be especially considered in treatment ; but the truth is, that not only the ejaculatory ducts, but all the extensive ramifica tions of the tubuli seminiferi, are in the same condition. The irritability even extends to those parts of the genitourinary system which have nothing to do with the semen ; as is familiarly known to practitioners. An irritable bladder, in fact, is the commonest symptom of sexual debility ; *raro mingitur castus*, the chaste, that is to say, the sound, pass water infrequently; while others have an equal want of retention of both fluids.

Nor the bladder exclusively, but the diseased action extends upwards, along the ureters, to the kidneys, disorganises those glands, which throw out a variety of deposits, as other affections are associated : such as the lithic acid, phosphate and oxolate of lime, lithato of ammonia, and others, easily distinguishable under a scientific analysis. I have kept a record of two hundred cases of deposit in the urine, occurring with spermatorrhœa ; and find that the oxolate of lime is the most frequent, in the proportion, within a fraction, of three to one. There may be simply mucus; or the urine may be quite transparent and watery. It is then more abundant, and passed frequently ; loses entirely its chymical qualities ; and abounds with saccharine matter, instead of uric acid and urea.

Masturbation is the only cause of weakness which I have hitherto adverted to ; which indeed deserves the foremost place, whether in point of frequency or injury. Many of the other causes are only auxiliary to this principal one ; unless we except excesses to Venus herself, and a too various worship. Promiscuous amours, which

c

are known to produce barrenness in the female, have an effect on the male also; and Nature punishes in that part which had offended against her. In whatever the custom of some countries has departed from the rule, the purpose of the Creator was, that man and woman should be of one flesh, and pair exclusively.

From indulgence, therefore, with the sex itself, whether with one woman or many, some of those effects may come, more commonly owing to abuses still more unnatural. Baron Larrey, that veteran surgeon who followed Napoleon in all his wars, from Egypt to Russia, attributes many instances of a wasted testicle in soldiers of the Imperial Guard, to excess of venery. Sir Benjamin Brodie has published a case. I might mention Sir Astley Cooper, and other authorities; but why multiply instances of an admitted fact; or step out of my own large experience, from which I could adduce, not an isolated case, but a hundred, if necessary. This is my exclusive domain in the profession; and where Sir Benjamin Brodie, for instance, in that large range of practice which his great knowledge and abilities have obtained for him, is consulted in one affair of this kind, I am consulted, as I have said, in a hundred, simply because it is my only business.

Since, therefore, it is known that, besides other effects, a wasting of the testis comes from over-gratification, it is next to be shown that from the very opposite cause the same effect may arise. I have had many devout young men consult me, who had entirely mastered their propensities, and never, this way or that, gave way to the sexual ardour. These consulted me for spermatorrhoea, some with a wasting of the testicle, which is only a result from the other affection. Most of these young men were brought up, and educated, among their sisters; in whose presence, as Hunter long ago remarked, this appetite entirely languishes. Some of these were sons of clergymen; some in orders; and all, I believe, from the country. Those who observe monastic rules are liable to this complaint, as might be expected; for the order of nature is the same, and on either side there is transgression.

Continence, under circumstances such as these, is voluntary; and the passions are permitted to cool in austerity and repose. But the situation may be reversed; and while every effort is allowed which can excite the appetite, it is denied the final purpose. I had a gentleman come up to me from Wales, sanguine complexion, and great muscularity; who mentioned that he had been affianced to a lady nearly two years, but that circumstances had delayed the marriage. There was the greatest confidence between the families; and he was in the habit of staying to a late hour, in company with his intended, after the others had retired. He did not conceal from me that he was kept in continual orgasm; that he was allowed for that long period of time, almost every night, every favour but the last; which occasioned him pains in the testicles and spermatic cord, nocturnal emissions, and afterwards a continual flow of semen. He had pains in his head, and sudden twitchings in the regions of the heart. Nor had he found separation sufficient for his cure; for he had been, with that intention, some months from home, before he applied to me. There was no difficulty, however, in the cure; for he had naturally great elasticity of constitution. I give this instance merely to exemplify the matter, for I have many apply under similar circumstances; or, at least, injured by a continued dalliance.

Another, and common, cause of debility in the generative organs is gonorrhœa, by which a single unfortunate amour may produce the bad results of a far greater number. The nature of this we shall better see hereafter, when we come to consider this peculiar sort of inflammation; how it may run along into the seminal tubes; or at least cause by sympathy a most violent disturbance of the testicle itself. Perhaps, however, I should rather say, that the gonorrhœa itself is less culpable than the medicines given injudiciously in the cure of it, such as turpentine, cantharides, cubeb pepper, and others.

Stricture, a consequence from gonorrhœa, by interrupting the course of urine and semen, is a perpetual irritation. It is the cause of disease all along the passages both urinary and spermatic; with so much the more danger as it is frequently unsuspected. The reader will find, in the third part of this work, an accurate description of the signs by which a stricture may be known; a very obscure disease, and occurring oftener than many imagine. The patient only knows it in its results, which are frequently

so distant and unlike the cause itself, that even the doctor
is at fault, and runs on a wrong inquiry.

A gentleman, forty years of age, of a very robust frame,
and powerful constitution, consulted me for seminal weakness. He had passed through the hands of several practitioners, who used all the ordinary means of cure, with
only a transitory advantage. He had vertigo, with palpitation, and was treated by a very eminent member of the
profession for disease of the heart. By all the stricture
had been overlooked, as it offered very little obstruction to
the urine, from a general dilatation of the urethra, which,
although narrower at that one point, had been distended
by the pressure of the urine, and thrown into pouches.
The account he gave me of the remedies he had used, and
manner of the treatment, disposed me to look for another
cause. The man was, besides, of so vigorous and athletic
a cast of body, that seminal weakness in him, I was convinced, could not be the primitive affection, but must have
been superinduced, and continued, by another. I thought
he might have a rupture, or perhaps engorgement of the
spermatic veins; but such was not the case. I then asked
him if he passed water freely; he replied, he did; but
there remained always a few drops in the urethra, which
afterwards escaped and wet the linen. My suspicions were
justified; a few drops of urine retained in this way, is one
of the known symptoms of stricture; I examined him;
found the obstruction, which was about half way down the
urethra; treated it accordingly; and saw all those remote
disturbances of seminal weakness, affections of the heart
and head, disappear with the cause which had induced
them.

When I mention stricture, gleet may be supposed
included, than which nothing more debilitates, not only
the spermatic system, but the body at large. It is wonderful, indeed, and proves abundantly the nature of irritation,
that one or two drops of fluid, limpid almost as water,
passed two or three times in the twenty-four hours from the
urethra, will produce a greater lassitude of the body, make
the mind more irritable, and more debilitate, than the loss
of a thousand times as much blood.

The next cause I shall set down is the indulgence of
wine, or ardent drink of any description; which, first ex-

citing the organs, brings reaction, in its turn, and depresses in proportion as it raised. I observe that cases of this kind are less common than formerly, when a bottle of Port after a dinner was thought a moderate allowance. Excess of food acts like excess of wine; especially when high seasoned, as with the curry-powder. There is in this country a fondness for highly-spiced soup, beyond what I have seen on the Continent, where soup is more the food of the people.

What excites is of one kind; what depresses another; all narcotics, therefore, such as tobacco, even used as snuff. It is some years since a Scotch gentleman, about fifty-five years of age, came up to town, on purpose to consult me for debility; who, finding some of his symptoms return after I had effected a thorough cure, came to complain to me; he was somewhat excited, and took out his snuff-box three or four times. I asked him how long he had been addicted to that habit; he replied, for several years, but more lately when in low spirits. I prescribed again for him, but bid him leave off the snuffing. He did so, with the result predicted; and is since convinced that it was the original cause of his impotence.

Some of the most difficult cases I have had to treat have been among the opium-eaters; for this powerful narcotic so much depresses the nervous energy, not simply of the generative parts, but all over the system, that it is an Herculean task to resuscitate it. A gentleman who had lost his wife, was in such extreme distress of mind, that he took laudanum for two or three nights, to obtain a little sleep. Finding himself restless one night that he did not take the dose, he had recourse to it again; and, by degrees, sometimes taking more, sometimes less, fell completely into the habit of it. In this way he continued for two years, when he married again; but his virile faculty, he discovered, was almost totally gone. He came to me, at once, and was not slow, of his own accord, to suspect the cause. He had resolution enough remaining to take no more of the drug; and was restored with less delay than might have been expected.

I must remark here, and shall have occasion most likely to remark again, that there is the greatest difference in the constitution of patients; some weakened by the slightest

causes; and some bearing the greatest errors of all sorts
with comparative impunity.   I have known a single act
of masturbation bring on spermatorrhœa; and I have
known others who had practised it for fifteen or twenty
years, before they had applied for assistance.   This observation will apply to all the other cases I am enumerating,
which I beg the reader will remember as he goes on.

Among all these causes, however, that which operates
the most gradually, is the force of a tropical climate.   No
one escapes its influence at last; and a great number of
my patients have been in the East or West Indies.   Some
correspond with me; a few come home on purpose; especially since the over-land route has been established.
Some are natives; who not less are overcome by a southern
sun, and show all the evidence of premature decay.   Heat
is a familiar cause of excitement, which induces a continual secretion of semen; which again the relaxed vessels
permit a passage to, and establish the spermatorrhœa.

Before I end this chapter, though I should pass over
many other causes prejudicial to health in the genital
function, one at least must be indicated, on account of its
great power and frequency.   In this I include all intense
application whether to books or business; which seems to
usurp upon the nervous energy, and take from all the
purely animal nature.   These natures, mental and corporeal, seem opposed to one another, and even in vulgar
opinion, as I might show by a common proverb.   The
lean figure and mortified physiognomy of the student
shew the general sluggishness of the circulation; but the
greatest atony is unseen, which lies in parts quite of
another nature to mind.   Yet it is very extraordinary
that they should operate so very powerfully one upon
the other; unless indeed it be admitted that a considerable part of the brain, which is the physical organ of the
mind, controls and is the spring itself of the venereal
impulse.   Gall and Spurzheim were not the first to suppose that the cerebellum was the seat of the amatory
power: and although the principles of phrenology have
been shaken, this fact remains, and is adopted into science.
We suppose, therefore, that when those parts of the brain,
or those of the mind properly, are kept in continual action,
they usurp upon the remainder, and diminish the pre-

ponderance of the animal part ; while, conversely, a great indulgence of this more animal function of it, debases the other, and levels the man with the brute.

There are not wanting many cases to prove the immediate relation of function between the testes, with the penis, and the cerebellum. Baron Larrey* gives the instance of a soldier struck in the back of the neck by a musket ball, which injured also the occiput. He recovered; but the testes wasted away, and the erectile power was destroyed in the penis. The same illustrious surgeon supplies another example ;† a man equally vigorous of constitution and passions, who lost from the wound of a sabre, the projecting part of the occiput, with injury to the dura mater underneath : he experienced sharp pains along the spine, and a tingling in the testes, which wasted, and, at the end of fifteen days, were not larger than a bean. All desire left him, and even remembrance, of the sexual indulgence.

A gentleman addressed me a note from his hotel, requesting to know at what hour I should be entirely disengaged to see him. Having made the appointment, a gentleman came in of a fine florid complexion, and broad shoulders, the picture of vigour. What was my surprise when told that he was very deficient ; and that he had scarcely any passion. He mentioned that he had been thrown from his horse while hunting ; that at first he had violent erections, and indulged immoderately with women. In a short time, however, the power left him as it were suddenly ; and he perceived the testicles wasted, and the penis retracted. Upon further inquiry, I learned that he had been thrown on his back, and received a concussion of the entire spine and back of the head. As there were still active symptoms, I applied leeches to the occiput, and used other means to lessen irritation in the cerebellum and medulla. This done, I resorted to such treatment as the case in other respects required ; and the gentleman was able to return home at the end of a few weeks, quite satisfied with the result. I will confess, however, that I am somewhat in doubt how far the wasting of the testes

* " Mémoirs de Chirurgie Militaire," p. 262.
† Loc. cit.

should be attributed to the fall, or to the excessive indulgence subsequently resorted to.

From the above instances, no doubt can be entertained that a large part of the brain, or that which is placed at the base posteriorly, superintends the sexual economy; and that, consequently, the inference I have drawn is easily explained, of the reciprocal relation between the mind in motion and the sexual sense. Perhaps, however, the phenomenon may be accounted for exclusively by that nervous interlacement and sympathy I have already enlarged upon : but every theory apart, the fact is undisputed, that excessive application of mind debilitates in this way, as so many men of business, so many professional men, and so many scholars, have experience of. Many of my patients are from either class; although the prevailing notion has been that scholars only bring on themselves this infirmity in this especial way :

"For love forsakes the breast where learning lies,
' And Venus sets ere Mercury can rise."

I may give the case of a gentleman, who, with continued application to business, had amassed a considerable fortune. He now thought of settling himself in life; for which he was young enough, not more than forty; and looked about accordingly for a suitable match; he found the daughter of a merchant in the city, with sufficient fortune, which he most wished for. He got married : but having long led a chaste life, with a busy one, he never had leisure for a certain matter; and now found that money is one thing, and enjoyment another. He came to me in great turmoil; and told me in a few expressions how he had failed. I found it necessary to examine the parts; the veins were much enlarged in the scrotum, like a collection of worms, and quite overlapped the testicle, which was not more than half the natural size. I have no doubt that the constant standing at the desk for upwards of twenty years, must have contributed to the enlargement of the veins. He had very little real passion. This, indeed, was the chief difficulty; and it was not until the veins were considerably strengthened, and the testicles developed accordingly, that the real sexual desire produced in him the usual effects of erection, with emission of semen in coitus. He is restored completely, and has several children.

CHAPTER III.

HAVING, in the preceding chapter, enumerated most of those causes inducing debility of the virile faculty, and especially the disease Spermatorrhœa; having also, incidentally, touched on the effects from such causes, which become so many symptoms of the disease; I will now disclose these effects universally, and give the whole subject to the reader in detail.

I have commonly made use of the expression "irritability" as a comprehensive phrase, to express that condition of the spermatic and urinary passages, brought on by whatever cause; it is not, however, to be concluded, that these passages always shew the same state, or differing only in degree. The remote effects, and all such impressions as are universal in the system, are, for all practical purposes, the same; but this unfortunately may lead to a great error as to the local cause of them, fatal in treatment, as it daily proves to be. The seat of the disease should be explored minutely; and the condition especially of the mucous lining of the urethra, ejaculatory ducts, vasa deferentia, and bladder.

Omitting varieties, which it must be remembered are numerous, there are three distinct species of morbid action in which these passages may be found, never to be confounded by the practitioner: 1st. there may be merely a relaxation of the mucous membrane, with disordered secretion, and feeble muscularity; 2nd. sub-acute inflammation; 3rd. ulceration of the membrane, leading ultimately to the entire destruction of these tubes, I mean the ejaculatory ducts. These states form a series, which may run into one another, as the disease advances; but they may, and do often, continue distinct; a patulous or inelastic condition of the membrane remaining for years, without assuming a more irritable condition; while, again, the chronic inflammation may continue on the same, unless some special provocation should promote to ulceration. However, when this last once occurs, complete obliteration is always to be apprehended.

Those who have examined the bodies of such as suffered under these several forms of complaint, describe the tubes, where they meet, as being of an unusal calibre, the lining membrane relaxed and loose in the tube, and paler than perhaps is natural; the bladder flaccid and capacious : this is the *first* form. . The *second* shews a contracted tube, with thickened coats, the internal of a roseate colour : the coats of the bladder much thickened, and the cavity contracted proportionally. The *third* an abraded surface; patches of ulceration ; or, even, entire absorption of the ejaculatory ducts, and adjoining part of the vasa deferentia. In all three, the kidneys shew a sympathetic condition: the testicles also are wasted; unless through congestion, from a want of tone in the vessels, especially the veins.

I will give the morbid anatomy in full of one case, which occurred thirty-three years ago; it was when I was yet a medical student; and I remember well to this day the terror it imprinted upon my mind, for I knew the young man, and had often associated with him. Some medical men still living will, doubtless, call the circumstances to mind, as they were much talked of, especially among the English students at that time in the French capital. Indeed there is one eminent physician now in London, then a young man, who knows all the details. My reason for dwelling especially on this case, is that it was the first cause of turning my attention to the subject, and determined me as to what part of my profession I should study especially ; in fact, I adopted this as a *spécialité*, and turned most of my studies in this direction.

R. E—— was of English parents, who were healthy, himself of a mixed sanguine and nervous temperament; and at the time he went over to Paris as a student in medicine about twenty-four years of age. He was a diligent student ; and, although of a good constitution, had weakened his system a little by a sedentary life and close application. He associated himself with one of those girls who abound in the *Quartier Latin*, known among the students as *les étudiantes.* I believe he was attached to the girl, more than is usual; but was greatly annoyed afterwards, and mortified, as she contracted a gonorrhœa, which she communicated to him. This lasted him almost a year; under an imprudent use of nitre, cantharides, and copaiba; and, in fact, laid the foundation of that irritation which ultimately ruined him.

Chiefly, I conclude, from chagrin at the infidelity of his mistress, he led afterwards a retired and ascetic kind of life, studying hard; but was known as an inveterate smoker; so that the young men used to jest with him, and, seeing him moisten the pipe with his lips, used to call it his *baiser d'amour.* In this way he went on for more than a year; but all this time indulged to a most incontinent degree in masturbation. Day or night, he confessed, did not stop him; and persevered in it with a gloomy resolution, although well aware of the fatal effects attending it. The first time I ever saw the work of Tissot *Sur l'Onanisme* was in his hands; a proof that he was not insensible of his condition, or the cause of it. He was always seen to lean on something; and even at the dissecting table, would support himself on his elbows. He had a short cough; and was very pale and emaciated. Symptoms of phthisis, in fine, developed themselves, which he paid little attention to; I believe he did not consult any one, with an entire disregard of all prudence. However, determined to bathe one afternoon in one of those machines constructed on the Seine, he superinduced more acute symptoms; called in, too late, one of the professors of the school, and died a few days afterwards.

A post-mortem examination disclosed the following: The brain was softened generally; with a slight serous effusion between the convolutions, in the ventricles, and at the base; patches of lymph were observed on the pia mater, over the left hemisphere; the arachnoid membrane was vascular, and adherent, on the left side, at two or three points, to the dura matter. In the chest, was found an effusion of blood and serum into the pleura, with recent adhesions; the lungs studded with miliary tubercules, distended with pus. Heart of natural appearance. No traces of disease in the stomach, liver, and intestines; kidney enlarged, lining membrane thickened and injected;

bladder contracted. Prostate twice the natural size, with the follicles considerably distended. The ejaculatory ducts were ulcerated at the opening; and much dilated and thickened. The septa of the vesiculæ were broken down. Higher up, the vas deferens was slender; the testicles were wasted, and the penis shrivelled. This concluded the examination. Many cases of this sort have been published, by different authors: but I prefer giving one which I had witnessed myself. Instances of the kind are very rare; at least, in my entire practice, having never lost a patient, I should say that this termination is seldom to be apprehended. This young man threw away his life, by sheer neglect, and a gloomy resolution; the more blameable in him as he might have commanded the advice of the most eminent physicians, attached to the hospitals and school of medicine in which he studied; in a capital where the subject has been much more considered, and is consequently much better understood, than it is generally amongst us.

Let not this instance, therefore, alarm any sensitive mind: remedies are at hand; and even the worst of those appearances just described are removable, under skilful treatment. I will here subjoin one case, which may be looked upon as the counterpart of the above; and which I select on account of the resemblance in all its symptoms, and in everything, except the termination. This was also a medical student, about the same age; introduced by his father, a surgeon of extensive practice in the country. He had passed a very abstemious life at home; but, on coming up to town, gave a loose to many irregularities. However, he would sometimes recollect himself; become thoughtful; and study very hard for weeks together. This made the young men think him odd and eccentric; who would combine, and try jests upon him, which at first 'he disregarded, but afterwards could not bear without pain. One evening they stole him off, and, as he suspected strongly, introduced him to a girl known to be diseased. Symptoms of gonorrhœa shewed themselves; but he was afraid to confess them, well knowing how much the joke would be at his expense. He treated the gonorrhœa in a clumsy way himself, from advice in books, and, in about nine or ten months, arrested it. But he took a thorough

dislike against his fellow-students, and as much against
girls of the town; confining himself after lecture to his
room; and studying with every perseverance.

At the close of the session, he went down to his father
in the country, who at once saw the great change in the
health of his son.   His eyes dim, complexion sallow, chest
contracted, and air listless.   The father suspected instantly the cause, and asked him if he had the venereal ;
which he denied, but acknowledged subsequently that he
had been cured of it.   It was difficult to get particulars
from him ; which made his father watch him more closely,
and inspect the sheets in which he slept.   These he found
covered with stains of a kind that could not be mistaken;
but the father made no remark, determined to spare a day,
and bring the young man up to town for advice.

I found him stooped of figure, drowsy, and nervous ;
voice husky ;   short cough, with expectoration ;   night
sweats ;   skin cold and clammy.   He mentioned, that
after dinner he was heated, particulary when he eat liber-
ally, which he was inclined to.   He had forgotten a good
deal of what he had learned, and found it difficult to recover it.   His father having left us together, I learned
from him some of those particulars I have already given ;
and also that, after the gonorrhœa, he addicted himself to
masturbation, as often as two or three times in the twenty-
four hours :   that he drank gin.   He had observed the
emissions carefully, which lately were streaked with blood,
and mixed with purulent matter.   I found the testicles
irritable ; with pain on pressure of the perineum, beneath
the prostate, which was enlarged.   These, with some
other symptoms, made it obvious that ulceration was
going on ;   and I proceeded without delay on that supposition.   The result proved the justness of the diagnosis;
the treatment succeeded; and all those other symptoms,
as the cough, depression, with loss of memory, and impaired virility, disappeared as if spontaneously.

I make not the least question that this case would have
run the same course as the other, a little longer neglected
in the same way; and I am convinced that a large number of those cases of consumption proving fatal in young
men, are induced by this cause.   At later periods of life,
ulceration in the urethra and ducts is less likely to super-

vene, for this reason : all the solid parts of the body are more firmly organised, and their irritability considerably lessened. Those in the middle stages of life, are likely to shew more chronic symptoms ; and, still later, mere relaxation, unless, indeed, with complication of other complaints. Enlargement of the prostate, which is a disease affecting nine-tenths of the old men in England, is an effect of the malady I am speaking of, and treated with a want of success inevitable, from overlooking the cause. With regard to the sequence, however, I must add, as much will depend on constitution as upon age. But with regard to constitutional symptoms of one class, it is in the period of life between thirty and fifty, that evidence will appear in the mind, deranged in every variety and degree, from restlessness, to melancholy and mania ; for the preponderance is more towards the brain, which in the young is towards the chest.

However, since no interval of life has an immunity from the evil effects I am adverting to, and as the subject must consequently be interesting to all, I will here trace it in that minute manner I have promised ; and commence with a description of the semen itself, without a proper knowledge of which, the simplest elements of the complaint cannot be comprehended.

Pure semen is only found in the testicle, and *vas deferens* ; for that emitted in coition is mixed with the several fluids supplied by the vesiculæ, the prostate, Cowper's glands, and lacunæ, which are certain slight depressions in the urethra. It is these fluids which impart the peculiar odour to it, which has been thought like that of the pollen shed by some plants ; for example, the chestnut-tree. Its specific gravity is greater than that of water, so that it sinks in it ; it liquifies spontaneously, but is not soluble in water. Under analysis, it is found to contain, of water 90 centimes, of animal mucilage 6, phosphate of lime 3, and of soda 1. It is the soda which gives it the alkaline reaction, and turns syrup of violets to green.

Examined, in a recent specimen, under the microscope, this fluid presents a strange animated appearance ; a number of creatures are seen to dart rapidly through it ; will be at rest an instant, and then are as busy as before. There can be no doubt that these are animalculæ, with regularly formed head and tail. The head, or larger part, is oval; the tail filiform; and the length, so diminutive are these creatures, not more than 1-50th of a line. The fluid in which they are seen to gambol is called the *liquor seminis;* which contains also a number of round corpuscles, or *seminal granules,* supposed to be the ova in which the spermatozoa are formed.

Fig. 13.*

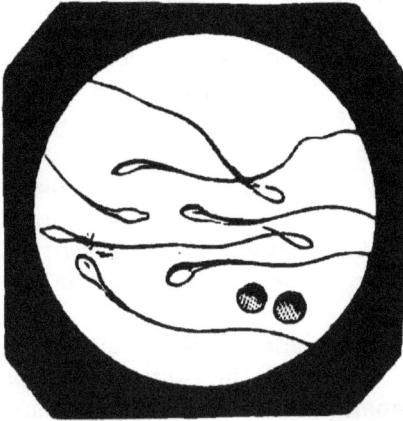

These creatures are very tenacious of vitality, and preserve their activity after the sexual act, for days in the vagina. They survive fully as long in blood, and in milk; but the saliva is obviously hurtful to them, as well as the urine. It has been said, that they perish instantly in the urine; but this is a mistake, as I can declare from repeated observation. The urine precipitates their motion at first, which continues for two or three hours; they dart convulsively across the field at intervals; and seem to die in torture. Cold water is equally pernicious to them; although one or two will remain in motion, after all the rest have perished.

Not only in the human sperm, but in that of all animals spermatozoa may be detected.

There is no longer any controversy whether these animalculæ are the essential prolific principle. They abound in birds at the time of pairing, and in the mamalia in the rutting season. They are not seen in man before the time of puberty; disappear in extreme old age; and are either excessively delicate, sluggish, or totally wanting, in the impotent. In the wasted testis they do not appear at all, and they are never seen in the mule.

In the prime of age and health, the quantity of semen thrown out in the sexual union, is from one to three

* Zoosperms, with Seminal Granules, magnified twelve hundred diameters.

drachms. But it is not so much the quantity, as the quality, which is the question; as a small quantity will produce all the sensation, and impregnate, with the same effect as a greater. Of the physical characters, consistence is the most desirable; for the semen is thin as first formed, and has its more fluid parts absorbed in its slow progress through the epididymis and vas deferens. When, therefore, in seminal weakness, the fluid comes out incontinently, it is with a diabetic fluidity. The sperm, in its proper consistency, is taken hold of, and thrown out, by the muscles, with so much the more energy; and sensation is in proportion lively. But the watery semen seems to escape the grasp of the muscular fibres; comes out in anticipation; while the accompanying sensations want the elasticity of delay, and the whole phenomenon is precipitated.

Generally speaking, in the vigorous the semen accumulates rapidly, and the desire, with the passion, to expel it, importunes so much oftener. I have had many patients, however, who explain that this was the commencement of the malady; who, in fact, by putting them too frequently in action, had irritated the parts. This irritation, again, exciting in the new, seemed but the natural impulse; which again and again obeyed, reflected on the cause, and thereby confirmed the malady.

So long, however, as the quantity of semen emitted approaches what may be considered the healthy, and that sensation is in the same degree vivid, the disease is less advanced. Nocturnal pullutions, however depressing, if accompanied with the natural excitement, are much less serious than those occuring listlessly. The worst of all is that atonic state, in which the semen comes after the urine at stool, or stains the linen in walking or riding. In a word, there are three stages of seminal weakness: nocturnal, with sensation, and without sensation; diurnal, which is completely passive.

If, from examination of the spermatic fluid, we turn to consider the urinary, we find a greater or less incontinence of it; the desire to micturate frequent, unexpected, sudden and imperious; with the jet failing at the close, and the last drops emitted without energy. The small muscles, seen in wood-cut Fig. 9, called acceleratores seminis, or urinæ, are so far paralyzed, as to fail in their proper action. The specific gravity of the fluid is low; falling to 1.008, assuming the healthy as at 1.020. This is the more usual; although in some cases of high coloured and turbid urine which I have examined, it has risen above 1.020, or the standard.

Let not the invalid conclude these instructions more minute than indispensable; I know the importance of them from experience; and will relate an instance to prove it. A bachelor, about the age of fifty, who had enjoyed his life, yet preserved a tolerably good constitution, experienced a very unwelcome failing within a year or two. The approaches were so gradual, that he suspected it to be the natural decay of years. Still he was not quite content to forego without an effort his former pastimes; and consulted one of the first physicians in London. Not deriving much satisfaction from the council or the remedies offered him, and having some notion of seminal discharges, with the method of detecting them, he requested that the urine might be examined. This was done accordingly, and I believe zealously enough; but without detection of the fluid suspected. The doctor pronounced positively that the animalculæ were not to be seen. General tonics were administered; which failing, after a while all treatment was discontinued. The patient, by no means at his ease under these circumstances, came to me, in turn. After some preliminary inquiries, I proposed to examine the urine; but he replied, that nothing could result from that investigation, as it had already been inspected. I learned in what way the specimen had been collected, which was carelessly from the chamber utensil. I therefore advised to collect the first few drops, passed on getting out of bed in the morning; and not only detected the zoosperms, but induced the patient himself to look through the glass, who saw them as plainly as I did. To ascertain the *cause*, is the first step towards the *cure:* I applied my remedies accordingly; and the result quite justified the prognosis.

Before turning to such results as are more general in the constitution, I must mention the peculiar appearance of the penis and scrotum; which in some habits are quite contracted; while in others these parts are much relaxed, the scrotum in particular, which is loose and pendulous. These conditions will depend partly on the greater or less irritability of the muscular fibre; but if the testicle is much wasted, it is more apt to be drawn near the body, and supported easily by the cremaster. However, if there is great relaxation of the tissues, such as comes from excesses or masturbation, it will hang loose, however light the testis, or diminished of volume.

The testis, it should be noticed, will often appear of a sufficient size, though its function is found to be much impaired. This occurs when there are depositions in the substance of the gland, when the epididymis is enlarged, or when the spermatic veins are distended and thickened; all which states are exceedingly injurious to the power of this part, and the common result of such abuses as I have so frequently mentioned.

Fig. 14.*

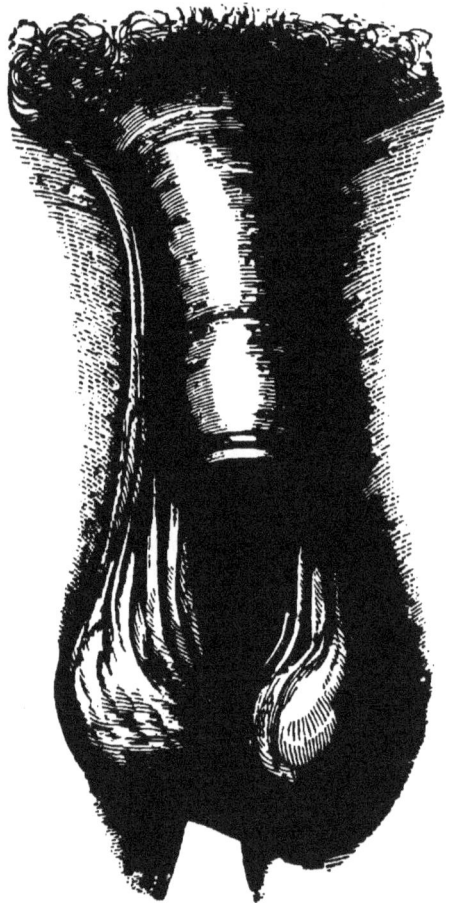

Such, therefore, as thus disclosed at length, being the local effects, we are now to investigate those of a more general description, and those more remote from the origin of the injury.

The greater number of those writers I quoted in the

* Relaxed Scrotum and Pendulous Testicle; the result of masturbation, and other excesses.

Historical Remarks dwell on a wasting of the flesh as a common symptom; who considering also the weakness of the back, give the disease the title of dorsal consumption, which is the expression of Hippocrates, φθίσις νωτίας. The countenance is pale, with languor and muscular relaxation. It is to be supposed, indeed, that from the drain of the constitution, and the usual results of irritation, the body should languish generally, with a sluggishness of every function. Yet must not this be stated without exception; for I have repeatedly treated patients for sexual infirmity, possessing every outward appearance of health and virility; hale, intrepid, strong of voice, and stout of heart; with only this deficiency. I account for this circumstance, either from great natural phlegm of constitution, resisting all irritation; or from such extraordinary activity in the digesting and assimilating powers, as to supply rapidly every waste from pollution.

Since the effect may extend generally over the frame, so it may determine in one direction, according to predisposition, established by habit or nature : indigestion, constipation, torpidity of liver, pains in the heart, spitting of blood, or neuralgia may, any of them, be the prominent result. The more is any disease likely to occur, as it has more of a nervous character, from the peculiar nature of the primitive irritation : affections of the heart, therefore, with disorder of the mucous, or nervous lining of the stomach; twitching of the muscles of the face; paralysis; defect of vision, from cataract, or disease of the retina, which is but an expansion of the optic nerve. The last mentioned is what we name amaurosis; the earliest evidence of which is an appearance of spots, or halos, floating before the eyes, called by surgeons *muscœ volitantes.* That eminent surgeon Mr. Travers has given, in his work on Diseases of the Eye, two cases of amaurosis ; the result of excesses, in one instance ; of masturbation, in the other. " The most pitiable cases of amaurosis," says Mr Travers, " are those of early life, from excess of sexual indulgence, and especially of solitary vices. The following are strong examples : A country lad, of robust constitution, became, alternately the favoured paramour of two females, his fellow servants, under the same roof. He was the object of gutta serena in less than a twelvemonth. Another at an early period of puberty, suddenly fell into a despondency, and shunned society. He never left his chamber except when the shades of night concealed him from observation, and then selected an unfrequented path. It was not discovered until too late, that in addition to other signs of nervous exhaustion, a palsy of the retina was the consequence of habitual masturbation."

While I am writing these pages, a case of this very kind is under my care; a gentleman who had consulted me for a recent gonorrhœa, and whom I remarked to be very imperfect of sight, with a dilated pupil. He called on me again, after his gonorrhœa was cured, to ask some question; when I was curious to inquire what was the matter with his eyes. He said that his eyes had failed him of late, from poring continually over books and papers, required by a situation he held under Government. I asked him bluntly, did he know what the word *masturbation* meant? He was confounded for a few seconds, and stammered out a reply. You must know, said I, that the practice I mention is a most serious cause of disorder in the eye. He confessed he had not heard it; and, afterwards, expressed his intense regret that the fact had not before been made known to him. It was by sheer chance, therefore, that he had not neglected the affection until "too late," as in Mr. Traver's case; and to the casual hint I threw out, owes his present improved, which I do not despair will soon be his perfect, vision.*

It is curious to remark that monkeys are very liable to cataract; caused, I should say, by masturbation, practised by those creatures in a most furious degree.

Consumption I have already adverted to as a sequence of spermatorrhœa. It is instructive to remark a difference of opinion among writers on this important subject; some asserting that consumptive patients retain their sexual appetite to the last; and others, that it fails, with a defect of semen, and absence of spermatozoa. Yet are these opinions in a great degree reconcileable; for he who has irritated his organs, and exhausted his frame, by excess ro masturbation, is not unlikely to continue his propensity to the close, and deceive with an appearance of the power, when it is only the disposition, to venery.

* Since the above was put into type, I am enabled to say that this gentleman, having first recovered the use of one eye, has now gained the use of the other also,

Instead of an inward determination, the blood may take more the direction of the face; much less alarming to life, but very inconvenient to such as mingle in society, and set a value on appearances. A crop of pustules spreads over the lips, cheeks, and nose, known technically as *acne*. This has been less suspected of its origin; but there is another description of invetrate eruption over the forehead, called the *Corona Veneris*, long recognised in its nature and its cause.

A frequent effect is chillness, and admitting of easy explanation. Animal heat is entirely a function of the nervous system; whatever, consequently, debilitates the nerves, must lower the power to resist the impression from cold, and keep up the circulation, particularly in the remoter parts.

"Those affected with this malady lose by little-and-little, like old men, the memory of facts, of dates, of numbers, and even of words; which increases still more their disposition to silence. After having commenced a sentence they often forget what they intend to say, or they can not find the expression they require; they confound themselves more and more, and finish by stammering, as if from a difficulty of articulation. This difficulty occurs in reality in the last stages of this malady; for the tongue is not then exempt from the disorder of the muscular system in general, and the irregularity of its movements is farther increased by the hesitation of ideas.

"Such persons forget continually their affairs, their promises, and their appointments; every thing, however important to them, as occurs to those who are falling into lunacy. Nevertheless, all these distractions do not depend exclusively on the loss of memory; the constant preoccupation of mind, and concentration of ideas on their complaint, make them inattentive to every thing else; or, when they neglect the most important matters, it is often because they take no interest in them."*

There is a constant apprehension of some accident or calamity; and, in a word, the entire train of hypochondriac affections follows on that nervous irritation, which is the beginning of all this various and complicated mischief.

This is of all maladies the most distressing; for not only is it without sympathy, it is treated with ridicule;

* Lallemand, tom. iii. p. 163.

and even the doctor himself endeavours to laugh his
patients out of it. Nothing can be more unphilosophical
than such conduct, or shewing more ignorance of the nature
of the affection. No one may be more sensible that many
of his sensations are fancied than the hypochondriac himself; but then he reflects that the fancy itself is not at
his control: there is the terror; he is aware of the hallu-
cination, and with great reason fears it will be worse.

I will select one case, out of a number, to illustrate this
matter. A gentleman consulted me, whom I found a
person of excellent understanding, of considerable learning
also, as sensible a man and as free from peculiarities as
any I ever spoke to. He simply complained of very disagreeable nervous symptoms, with local relaxation, caused
by youthful indiscretions. I treated the case in the
usual way; and it was when he began to improve that he
ventured to explain what he had meant by the nervousness he had complained of. He declared that he wanted
courage to express himself before; and that even now he
could not look back on his previous state of mind without
shuddering. He had fancied his head many times the
natural size, but the remainder of the body dwindled away
to that of a pigmy. " I knew this," said he, " to be a
mere error of sensation; but notwithstanding I could not
pass through at a door without a painful effort, lest I
should strike against it." "Of course," he added, " I
knew the door was large enough, and that it was a mere
fancy; but I was oppressed with the apprehension lest the
mind should fail altogether, and that I should be put
under constraint."

M. Lallemand has a passage quite to the present purpose, which I will make no apology for transcribing. I
will only remark, that this is the most excellent and
valuable part of his work; probably from his studies having been previously directed to disorders of the mind in
general. " It is in vain," says he, " that we say to the
so-called hypochondriac, amuse yourself, employ your
mind, go into society, seek agreeable conversations: so
long as we have not removed the cause of his disorder, he
is unable to profit by our counsels. How can we expect
that when a man is fatigued by the least exercise, he shall
occupy himself with walking, or gardening? How can
we desire him to go into society, when the simple presence

of a woman intimidates him, and recalls all his former misfortunes? How can we expect him to enjoy conversation, when he loses its thread every moment; when his memory fails him, and when he feels his nullity? We ruade him to seek amusements and pleasures; but are they such to him? Is not the happiness of others his greatest punishment? Because he is unable to follow our advice we accuse him of unwillingness, and we wish to compel him. Let us first remove the cause of our patient's disease, and we shall soon see that his character and conduct will change, and that he will return to his natural tastes and habits."

In this suggestion, undoubtedly, is the true method; let us bring the matter to a practical issue; remove the supposed cause, and decide upon it by the results which follow. If we find that variety of nervous affections which we include in the term hypochondria, to disappear with the spermatic irritation, we are forced to conclude, not indeed that spermatorrhœa is hypochondria, but that it promoted and continued it. So many cases have passed through my own hands to justify this mode of inference, that I am rather at a loss how to select amongst them; and fearing that in advancing one or two cases, the reader should think them exceptions, and not examples from a great number. However, I will give an instance, which I select rather because the symptoms are of a more moderate kind, and such as are most common; unwilling to shock the reader by an extreme instance, with the total perversion of intellect.

A gentleman who had been employed in some negotiations in the southern part of Europe, tendered his resignation unexpectedly, and came home to England. He was, I suppose, forty-five years of age; portly frame; complexion sallow; and, when I first saw him, saturnine, absent, and melancholy. He told me, after some days, when he had become more familiar, for at first he was altogether reserved, that he saw a great change in England since his return; that Ministers had not used him well; that he was the common mark of people's observation in the streets; and that, in society, his friends would whisper, and look at him; that he was forced into solitude by this impertinence, although very unwillingly. He had no doubt that this usage had seriously altered him; for he had remarked even his servants turn back, on leaving the room, and look at him. Sometimes, he said, he rallied; disregarded all remarks; and would hold himself perfectly indifferent, if not troubled by loud noises in his head, and something like the cracking of a whip, apparently in the substance of the brain. He had some knowledge of anatomy; and feared organic changes in the sensorium itself; adding, that he should fall a victim to uneasiness of mind, caused by those he had served too faithfully.

I had to use him with a good deal of delicacy; for he was alive to every suspicion; and I could hardly ask him how he did, without his asking me immediately, what I meant by that inquiry. However, it was absolutely necessary to get particulars, in order to form a diagnosis; and I made bold to ask him, what kind of a life he had led while abroad; for he had been in very good health formerly. He said he had lived temperately, but had taken less exercise, especially on horseback, than he had been acustomed to. I then told him plainly what I wished to come at; and asked him how he had conducted himself in respect of women. " Well, to be plain," said he, " I was somewhat captivated by those foreigners; and partly from excitement of climate, partly from novelty, I indulged much more in that way than I had done previously." He added that he had sometimes roused his jaded power, by artificial means, and taken especially some lozenges, obtainable in the *botica y drogueria* for that purpose.

Having procured these facts, I next got a specimen of the urine, (obtained in that way already described in this work;) and submitting it to the microscope, instantly distinguished the zoosperms, which were unusually abundant. I lost no time to inform him of this discovery; explained to him fully in what way all his symptoms could proceed from the seminal flux; and at length convinced him : though not indeed before he saw the vapours of his mind clear off, and his good humour return, under the treatment adopted.

What the course of treatment was I will soon give the reader to understand; and will now enter without further delay on this part of my subject, including all that relates to the most effectual means of relieving every form and effect of seminal irritation, coming from every variety of cause.

CHAPTER IV.

ON THE TREATMENT OF IMPOTENCE, WITH PHYSICAL EXHAUSTION IN GENERAL, AND DEPRESSION OF MIND.

THE reader who has been attentive enough to study the foregoing chapters, will not consider them superfluous to the result desired, which is the treatment and cure of spermatic affections, if he considers that all I have said of effects is but a description of so many symptoms by which disease may be recognized, and that what I have enumerated as causes becomes an immediate part of treatment by indicating what is to be avoided. Thus, for instance, the gentleman I treated for cataract did not know that this affection was the result of solitary vice; and therefore would not have discontinued it, leaving an insuperable bar to his recovery.

Let the reader then begin his endeavours at a cure by reconsidering the preceding pages, and thereby learn in the first place what he should attribute his present uneasiness to. Has he a weakness of the loins, a delicate stomach, an occasional cough, palpitation of the heart, an eruption on the face; is his memory bad; has he a singing in his ears, noises in his head, slight pain in the temples, with want of sleep and broken dreams by night, and by day weariness, melancholy, and distraction? If he has any of these unsuspected of its origin, let him next inquire of himself what he can attribute it to; and, wanting any other satisfactory solution, let him resolve whether he has ever trespassed in the manner of Onan, or whether he has bent too frequently before the shrine of Venus, or offended her by neglect.

Satisfied in this way of the nature of his complaint, he has at the same time got a knowledge of the cause of it; and his first purpose is to discontinue this cause, however much habit may have confirmed it upon him. Most persons have resolution enough to stop that one solitary shame, which is the gigantic evil of these times, and leaves every other cause of sexual infirmity in the shade. Others are more wedded to this harlot of their own creation; and though she disgusts to-day, offers again with meretricious seducement to-morrow. Such has been the infatuation of some unfortunate persons, that they knew no other way to be rid of this artificial passion than by removing the part to be gratified. "A man,* aged twenty-two, was brought to the London Hospital in January, 1836, having cut out both his testicles. · He had removed a small piece of the integuments, and squeezed the testicles out through the opening and excised them, having previously tied a piece of string tightly round the spermatic cords to restrain the hæmorrhage. These had retracted into the inguinal canals; and Mr. Adams, who was called to the case, was compelled to introduce his fingers at the wound and draw down the cords, in order to secure the vessels separately. The man admitted that he had been in the habit of constantly practising masturbation; and it was to rid himself of the perpetual desire to commit what he regarded as a great sin that he determined to remove the testicles." This is by no means a solitary instance; many others, like the pious Origen, have castrated themselves from this motive, or, as a boy in Edinburgh expressed it to Mr. Liston, that he might lead "a holy life."

I will take another instance from Mr. Curling's work:—
"I have been informed," says he, "by a professional friend, of a case in which double castration was performed, at the urgent request of the patient, on account of most distressing self-pollutions, that had a very lamentable result. The patient, a gentleman in the upper ranks of life, committed suicide; and the surgeon, who had been rash enough to emasculate him, was threatened by the patient's friends with an action at law for performing so unwarrantable an operation."

Having therefore laid it down as the first grand rule, to discontinue the cause, next we are to proceed to the less negative part of the treatment, for which I will give directions fully as precise, but not quite so compendious. I hope the reader has too much sense and knowledge to imagine that, with the charlatan of the East, I can say, "Here, take this potion, no matter what the form of your disease may be, it will infallibly cure it." My method, I confess, is more philosophical; I do not treat the old and

---

* Curling on the Testis, p. 116.

young alike, give the same remedies for an irritable urethra and a callous one, promote by the same dose the flow of urine when deficient, and arrest it when it comes in too great quantity. Reason and experience alike indicate that some discrimination is necessary, and that it is as blind quackery and imposture to pretend in these days to any elixir, as it once was to boast the philosopher's stone.

Classification I am sensible is usually very tedious to readers, which yet assists the memory, and is absolutely indispensable to understand and treat disease of any kind. If the reader, therefore, is also the patient, let him be attentive to the directions I will here give him, unless he would dissipate his efforts in blunder, and terminate them in disappointment. There are three essential varieties of the disease in question (as amply explained already), requiring a different treatment for each, at least in the first part, for afterwards there is more uniformity. There may simply be relaxation and want of tone in the parts ; next, irritability and sub-acute inflammation ; or, which is the most rare, ulceration, threatening total disorganization. Of each of these in its turn.

It is not difficult to ascertain when the sexual organs want sufficient vitality and aptitude of impressions. The scrotum is relaxed, the penis flabby, the glands pale, and little sensibility in the urethra. While there is an incontinence of urine in cold weather, it is retained longer in summer, and is much less of quantity. Pollutions, should they come, are more likely in warm damp weather, and are stopped by dry winds and a sharp air.

As soon as we are sufficiently apprized by symptoms like these of the nature of the affection, we should begin by the use of external applications, of which the most simple is cold. Cold is a powerful agent, but, like every other great power, requires a skilful hand to conduct it. Its first effect on the human body is to depress the vital powers, and if followed up, entirely benumbs, and extinguishes the lamp of vitality. In proportion to the strength of the frame is its power to resist this influence, and this should be our guide in making use of it. Every vital function, whether in the body generally or in a part, feeling the shock of cold, takes a sort of alarm, and reacts upon it instantly. In the strong this reaction is more lasting, and the operation suffered longer without inconvenience; but in the debilitated a watery impression is scarce borne, and the troubled effort of the constitution to reassert itself is shown by rigors. We always, in fact, have a clue to the aptitude of cold in any instance; if there is a general glow over the frame, or the part, with an agreeable feeling, the purpose has been answered; but if the body remains pale, or the parts constricted, we may be assured that the remedy is inapplicable, or that it has been injudiciously made use of.

Some caution is here necessary with regard to the use of cold, which, in the winter, is entirely inadmissible. In that rigorous season the system can scarcely maintain its own heat, much less resist any adventitious cause to remove it. Before resorting to this agent at all, the stomach and alimentary system must be considered; the effects on the head closely observed; subsequently, the state of the kidneys, and the local effects on the genitals. These are particulars which the practised eye of the physician can alone be judge of, and which, therefore, can never be dispensed with. The same agent, in fact, may produce the most beneficial results, or be greatly injurious, as it is made use of with judgment or injudiciously. I have had patients come to me with organic disease of the heart, brought on by the blundering use of the shower-bath. It would take up too much space, or I could give, from my notes, many cases of injury to the sexual parts from over-stimulation by cold. But, in addition to these cautions, the chief one is, that the urethra may be irritable or inflamed, the kidneys relaxed, and, therefore, this whole treatment contra-indicated.

I was consulted by a gentleman who complained of continued pains in the loins, and great prostration of strength in general. He drew my attention to the state of his urine, which threw down a heavy deposit of oxalate of lime, mixed with mucus. It was easy to discover that the kidneys were considerably involved, with the entire mucous lining of the bladder and urethra. Among other parts of the treatment, I deemed it essential to wrap the loins in flannel, and upwards to the dorsal spine. He asked me if he should discontinue the cold hip-bath he was in the habit of using night and morning. Certainly,

I said ; what have you been using it for ? He then told me, that previous to his marriage, about six months before, he was advised by a medical man to use the cold hip-bath in this way, with a view to increase his strength. I instructed him to give it over at once, and that he himself should judge of the impropriety of it. I also recommended separation for awhile from his wife ; and to give up all idea for the present of strengthening his system in the sense he explained. There was extreme torpidity of the bowels, from local irritation; and, although he ultimately recovered, I was apprehensive for some time that his life could not be saved.

When, however, by a skilful inquiry into the actual state of the spermatic and urinary passages, we are assured that there is simple atony (for what error more fatal than to confound this state with irritability), many other means are within our reach, some applicable to one case, some to another, according to those peculiarities distinguishable only by experience. I will just allude to friction of the loins, scrotum, thighs, and penis, with aromatic spirituous embrocations, as subsidiary resources.

In these atonic cases also without irritation, another powerful agent is galvanism, which stimulates more than any other we are acquainted with the nervous influence itself. The reader is, no doubt, aware that if in the dead body of any animal, the nerves and muscles be exposed, and brought into contact with the opposite poles of the Voltaic apparatus, the muscles contract, and move the parts they are attached to, with every appearance as if in life. On the first discovery of this principle, it was supposed that the real nature of the nervous or vital power itself had been discovered ; so nearly did it approximate to the living action. Certain it is, that when the nervous current is interrupted in a part, or generally in the body, galvanism is the most efficient means to re-establish it ; as seen so often employed in cases of suspended animation from submersion, or any other cause. It is analogous to that other, or negative agent, cold, in one respect, that what stimulates in a moderate appliance, in excess, destroys life altogether.

Electricity is another power, with effects very much resembling those of galvanism ; and it has been found from experience that a union of the two is most beneficial, when the nerves and muscles have been weakened in any part or organ. When, therefore, the practitioner is fully satisfied that the case is one adapted to it, the electro-galvanic current should be sent from the origin to the distribution of the nervous streams, taking care not to overcharge the machine, or push it so far as to occasion irritation, when it becomes absolutely hurtful. The shocks should be passed from the source of the spermatic and pudic nerves, as high as the renal, and through the lumbar plexus, down to that central point of the perineum where the smaller muscles are attached.

I will give one case more, at the risk of being thought tedious, to expose the vital error of remedial agents out of place ; for there is hardly a valuable therapeutic means which may not be compromised by the indiscriminate employment of it. A gentleman, somewhat advanced in life, mentioned to me that he had very little confidence in treatment, although he was willing to give it a trial. He was in the habit of indulging in wine, which I prohibited; and, indeed, found it necessary to put him on a course of treatment calculated more for an irritable than a relaxed condition of the seminal and urinary passages.

This done, after an interval of a few weeks, as a part of additional treatment, I offered to use the electro-galvanic apparatus, but he said it had been already tried, with no benefit whatever. I asked him to excuse me, remarking that I believed the circumstances were very different. I did not explain to him what perhaps he could not fully comprehend, the difference between irritability and atony ; but I satisfied him by the *result*, that the same remedy which proved injurious at one time, might be equally beneficial at another.

While these, or other, local applications are being used, internal remedies must co-operate with them, calculated to produce more permanent effects, and recover the lost tonicity of the tissues. These remedies I will mention presently; but think it more advisable, in the first place, to dwell on the other, or irritable condition, when, not only these, but all those powers of cold, galvanism, and electricity, in the first stage of the treatment manifestly injurious, may be employed subsequently with as decided advantage.

As it was not difficult to pronounce on the state of simple tonicity, so it is equally easy to recognise that irritable condition which is the second form of spermatic disorder. This may discover, in several degrees, from simple excitement, to absolute inflammation. In the simple excited state, the urine is increased in quantity, and paler, if it be in winter ; less copious, and probably clouded, in warm weather ; the first jet thrown hastily, the last drops descending feebly. The glans and orifice of the urethra may show nothing unusual ; but if the orifice of the urethra is redder than common, and that other symptoms, such as uneasiness in the perineum, and heaviness in the loins, draw attention also, we know for a certainty that there is actual irritation, or subacute inflammation of the mucous lining. When, in addition, there is evidence of involuntary nocturnal or diurnal seminal pollutions, we are infallibly assured that this state of the lining membrane is continued all along to the utmost convolution of vasa deferentia and tubuli.

This is the more usual condition in which patients present themselves ; nine-tenths, at least, are of this sort ; yet, the condition is such, that all those means hitherto mentioned are, far from being suitable, strictly to be prohibited. Whatever so operates on a part endowed with vitality as to produce reaction, must not be thought of when the part is inflamed, or in a state approaching to it. If there should be a temporary appearance of vigour, the subsequent increased vascularity is thereby only rendered more difficult of reduction. It has been the common error, even among the highest of the profession, to confound these states, and to use those remedies, suitable only to the few, to all cases whatever, with undistinguishing perseverance. For this reason it is, that you read in medical works, that there are few aphrodisiacs, temporary in their effects, and disappointing totally. They were administered at an improper time ; just with as fatal effect as when it was the rule with practitioners to administer bark and wine in typhus, without regard to that inflammatory state which we now know well must be first subdued before any tonic or stimulant is to be thought of.

The means to reduce the inflamed, or irritable state of the lining, which I would also call the nervous, membrane

of the urinary and spermatic passages, are in the first
place to be resorted to. All stimulants must be avoided;
exercise denied, especially on horseback : food simple, or
the milk diet, as recommended by Hippocrates, but with
a total absence of spices ; coition especially not to be attempted : and, indeed, the presence of whatever excites,
society of women, and reading of amorous works, to be
strictly prohibited :—

"Parce tamen Veneri, mollesque, ante omnia, vita
Concubitus, nihil est nocuum magis."

Still more must tonic remedies not be thought of, and all
that class of ferruginous and balsamic medicines exhibited
on the supposition that there is simply relaxation. I am
convinced that from ill-judged efforts to force, by direct
stimulants, the weakened organs, more persons have been
rendered hopelessly impotent than from any accidental or
other causes whatever. So necessary is it to be careful at
the outset ; and, as I am fond of expressing it, to know
the cause, and the nature of the malady, before we begin
the cure of it.

The internal remedies may all be included in one expression, those which diminish diseased action without
reducing power; which remove the irritable or inflamed
condition of the parts, but leave their strength unimpaired.
The time of using these remedies, the doses, and other
indications, altering with peculiarity of constitution, age,
temperament, previous treatment, duration of the malady,
and actual state, can only be given when particulars are
explained. At the close of the volume will be found minute
directions for this purpose, to which I must refer the reader.

When, therefore, by these, and like appliances, we have
succeeded in our first grand object of reducing all vascular
and nervous irritability in the urethra, seminal tubes, and
more especially ejaculatory ducts, then it is that the more
invigorating method is to come in turn, but not over-hastily,
and rather by approaches, until the parts have become
quite corroborated. Too much haste might throw us all
back again, and make the entire process more tedious than
before. It is better for the patient to lose a week or ten
days in those preliminary measures, than, by attempting to

force nature, cause her to resist, and set herself against any artificial interference. It is incalculable what mischief has been done by the neglect of this simple precaution, arising from the overhaste of people to restore in a night or two, as it were, or perhaps by a single dose, an affection, that has been the growth of years. From this abundant error it is, that cantharides, phosphorus, with other articles of the Materia Medica, have been so much resorted to by quacks, and have been so ruinous to their dupes.

The proper strengthening and stimulating medicines are very different from these, each in its proper place. I am in the habit of administering a preparation of the ergot of rye with marked advantage. Also the Cannabis Indica. Another is the nux vomica which, in combination, is more adapted to other cases. Among this class are the oleo-resins, gradually advanced from small doses; tar-water, and the warmer gums in general. The preparations of iron are inadmissable in every case, unless it be the lactate, in small doses; but with this precaution, that there is no irritation of the liver or lining membrane of the stomach and duodenum.

Dr. Johnson used to say that the ultimate object of every man, no matter how various his pursuits or ambition, was to be happy at home. But not to dwell on the obvious relations of man or woman in society, the obligations of morals, and that retreat from the distractions of life man seeks in the bosom of a family, his enjoyment in youth, his solace in age; there is the physiological consideration that the semen, like all other secretions, accumulating, requires an outlet, and is retained with injury. To solicit artificially, or to constrain it of necessity in nocturnal emissions, are each of them prejudicial, and not the original purpose, which was to increase and multiply, the most universal of all laws in nature, superior to that of defending life itself, as I explained in the Introduction to this volume. The advice of marriage, indeed, may with propriety be advanced as subsidiary to other treatment, sometimes preventing a relapse either from abuse of the instincts of nature, or too much inattention to them.

Sometimes, however, marriage, which may be used to confirm the cure, may itself be the cause of the malady, or, at least, one of the causes. We should bear in mind the dictum of Hippocrates, that this disease occurs in the *newly-married* and libertines. This species of excess, indeed, was almost the only cause known to antiquity, of what has been since styled Tabes Dorsalis; those other causes, so numerous as we have seen them, not known, or not adverted to. The inference is obvious; and yet, through a want of reflection the most unaccountable, it is the commonest thing with practitioners to recommend marriage to those suffering from spermatorrhœa.

A gentleman, about thirty-five years of age, put himself under my care, who had lived on his estate in the country, which he had always undertaken the management of himself; who got up early, was abroad all day, and retired soon, with uninterrupted good health. He never contracted any disease; and lived as free from any circumstances that could irritate or weaken the procreative propensity as any one that has ever consulted me. I need not run into particulars; this gentleman got married; and consulted me six months subsequently, complaining of a general lassitude. He had been from home some weeks, on pretence of business, but in reality to recruit his strength, by travelling; and although better as to his general health, he had so much trouble from pains in the loins especially, that he wished to have medical advice.

There was much tenderness of the testicles; and the semen oozed continually from the ejaculatory ducts, which were much irritated. It was incredible the quantity that came away in twenty-four hours. As the semen was never retained in its receptacles, there was not enough of it for copulation; erection was of short duration, and the fluid anticipated the orgasm. Here was a man, in fact, who brought himself into a very serious condition, not unlike that into which a more criminal indulgence precipitates so many. The case resulted favourably; and I here bring it forward to illustrate an important fact, yet very generally overlooked.

Let this much suffice for the first and second varieties of sexual infirmity :—There remains but the third, which is by far the most rare, as the preceding is the most common: happy for mankind, that, in their various calamities, the worst, always apprehended, yet seldom comes. I have hitherto kept no disguise with the reader, and I will in this instance also be equally candid with him. The treatment in this form is in a great measure surgical, and such as requires a personal application. I will not do injustice to a description the merits of which are in a long detail by giving the disjointed parts of it, and must advise the patient to look out for suitable advice. If he has arrived at this advanced stage, or the organic derangement of the seminal passages, (such as he will have found described in chap. iii. p. 47), his case is indeed serious, and he can have no thought of conducting the treatment himself. Indeed, in any case it is preferable to have the advice of a physician, for no one is a fit judge of his own sensations ; and even doctors themselves, as is well known, have a distrust in their own judgment when sick, and prefer taking the opinion of a brother in the profession.

However, those who by reason of distance cannot obtain, or those who from motives of delicacy do not desire, an interview, may have the treatment conducted by correspondence, and prescriptions or medicine forwarded to the address. Correspondence is a laborious and less satisfactory method ; and those who resort to it must not expect, unless in more moderate cases, the same rapidity of progress as under immediate superintendence of the physician. The result can certainly be obtained ; and those who wish to consult by letter had better describe the symptoms, whether mental or corporeal, in the simplest expressions, without any technicalities. The usual consulting fee of a respectable physician is a guinea. I have cured hundreds through the intervention of letters only, whom I have never seen, and whose real names very likely I do not know. Some have obviously used a feigned hand-writing ; a thing entirely unnecessary, as I destroy all such correspondence when I have taken memoranda of the substance of it. Many request to be addressed, "Post office, till called for;" and use other methods of concealment as it answers their purpose. Many are the grateful expressions and compliments sent to me ; but let not those who have written them be in terror that I shall print them, as *others do*, for the mere purposes of puffery.

Such, therefore, as please may have recourse to these or other methods of intercourse ; but when it is only false delicacy that stands in the way, that is entirely out of place with a physician, especially one in whom long familiarity has destroyed all impertinent curiosity, and who looks at the subject in a mere professional light. I am always consulted *alone* and *in person*, so that by no chance is there a possibility of exposure. The special nature of my practice keeps me isolated, and in a peculiar position in society; so that I may be said to be at once known, and not known. I do not expect that persons of rank and others who have derived benefit from my advice, should recognise me afterwards, however familiar, generous, and grateful with me in private. This lies altogether with themselves; and I have the patronage of those whose names shall for ever remain concealed, and whom to know is an honour.

☞ See Instructions to Patients at the end of the volume.

PART THE THIRD.

ON THE VENEREAL DISEASE.

———◆◆———

INTRODUCTION.

THE Venereal Disease occurs in a great variety of forms, all of which I shall describe. There are, however, two familiar distinctions; the Gonorrhœa, which is a running from the urethra; and Chancre, which is an ulcer, appearing usually on the penis, and which, with its effects, is called Syphilis.

This running and ulcer I mention, however inconvenient, painful, and dangerous, are yet not properly the disease itself, but the first impression of it, from which the remaining parts of the body subsequently become affected. The chancre, especially, is the chief inlet to the disease; from which it spreads, and contaminates universally: it is, in fact, a small receptacle of poison, thence absorbed, and producing peculiar effects, more. extended, more durable, and injurious. As the primitive ulcer is called the *primary* affection, these ulterior effects are called *secondary;* to which has been added a *tertiary* class, occurring in the same succession.

GONORRHŒA.

CHAPTER I.

THE SYMPTOMS OF GONORRHŒA.

THE first evidence of this disease is a slight itching at the extremity of the penis, and in the course of the urethra. The opening, if closely examined, will be found somewhat redder than usual, and the sides of it adhering a day or two later. These are the earliest symptoms of inflammation; and the gluing of the opening is from the matter just commencing, which had dried.

A little later still, if the finger be pressed forward from the scrotum along the lower surface of the penis, a few drops of thick yellow matter may be pressed out. At this period the organ is redder than before, and there is a scalding on passing the water.

As the disease progresses, the matter flows more abundantly, with a deeper yellow, and the scalding is more severe. The last drops of urine come away with difficulty, and a very acute sensation is felt lower down, behind the scrotum. This comes from the spasm of the muscles that propel that fluid, which compress, and thereby injure, the lining membrane of the urethra, made more tender by inflammation. A few drops of blood may follow on this injury; but only in more violent cases.

At this period, also, the glans is livid, and the whole body of the penis enlarged and distended. Erections consequently follow, which are excessive, and irritate the urethra by over-extension. However, the organ sometimes cannot fully unfold itself, held on one side as if by a cord, with much pain and inconvenience. This affection is named *chordée*, and is caused by adhesion of the cells of the penis, which will not admit of blood at this part.

After some time longer, the matter becomes more fluid, the desire to pass water more frequent and irresistible; the irritation of the penis in some degree subsides, and that of the bladder increases proportionally; the running becomes less, and may be seen even to have ceased; but there is a heavy sensation in the loins, thirst, and a general depression of the spirits. However, another revolution is at hand : these new symptoms, after an uncertain interval, subside, and the whole violence of the disease is transferred again to its primary place, with a renewal of the running; to the dismay of the sufferer, who supposed himself getting well of his own accord. Something of this kind may re-occur many times, with alternate importunity, between bladder and urethra.

Such is the most accurate general description that can be given of this complaint at the outset; its course left to nature, without any interruption from medicine. Treatment, good or bad, may alter, one way or the other, this succession of appearances; the same may be said of constitntion, which alters the character of all complaints.

No circumstance, however, can explain the more or less rapid development of the disease; which sometimes I have noticed in patients within twelve hours after the debauch, and which in a few instances I have known full three months later. The captain of a ship once consulted me, who had been upwards of ninety days at sea; when at last an acute running came on : there was no woman on board; all the crew were well; so that the seeds of the gonorrhœa must have remained lurking in the urethra from the night before he sailed, when he took a girl from the street.

CHRONIC GONORRHŒA.

When the gonorrhœa has lasted for some time, (I mean, left to itself, without treatment), the inflammation subsides of its own accord; and, after some interruptions, settles into what is called the *chronic* state, from the Greek word χρονος, importing *duration*. The pain subsides, the discharge is less like pus than mucus, and much smaller in quantity. There is a peculiar appearance; a little white thread, which may be drawn out from the urethra, tough, and slightly elastic. There is no violent erection, or chordee. The inexperienced imagine the disease subsiding, which, on the contrary, is much more tedious, even to last for years, if not checked by proper remedies. In this state, it is frequently called the gleet : but this is a mistake, into which not only patients but authors have fallen ; for the gleet has characters appropriate to itself, which I shall next particularize.

GLEET.

Gleet is a thin watery discharge, scarcely seen in the urethra, but staining the linen, as if with a solution of gum. There is much lassitude and weariness ; pain in the loins ; headache ; general debility ; but there is also a local debility ; for the organ appears pale and flaccid ; the urine dribbles away ; and the venereal ability is impaired. Gleet, in a word, is a frequent cause of impotence, as I have already shown in another part of this work.

EXTERNAL GONORRHŒA.

Gonorrhœa, usually in the place I have described, may have its seat externally, instead of in the urethra : I mean between the foreskin and head of the penis, then styled *preputial gonorrhœa* or *balanitis*.

EFFECTS OF GONORRHŒA.

Having thus made the reader familiar with the external or spurious, as well as with the internal or real gonorrhœa, I will now trace some of the concomitant effects of gonorrhœa, such as it occurs in the urethra.

SWELLED TESTICLE.

The first of these effects I shall notice is the swelling of the testicle, as it is one of the most common. The patient at first, the gonorrhœa yet recent and in the inflammatory stage, feels a sensation of heaviness in the scrotum, where the veins are somewhat distended. He perhaps has walked too much that day, gone on horseback, or in a carriage, or given the part a slight injury in sitting. By a little rest these symptoms may recede ; but more usually the testicle becomes enlarged, redder, and painful ; its own weight is too much, and requires support from the hand.

If not timely treated, the swelling increases enormously; and the heat is so great, that the patient is glad to cool the part with his breath. The body sympathizes; and there is thirst and fever, with sickness of the stomach. So serious a disease might naturally give serious apprehensions; but it is only through injudicious treatment, or neglect, that it becomes so formidable; and violent as it appears, is easily reduced and curable under more skilful management.

STRICTURE.

If we consider that so narrow a tube as the urethra is near a foot in length, it will not give surprise that such violent inflammation, with so much pain and discharge of matter, as come in gonorrhœa, should sometimes narrow it still more, and obstruct the flow of urine. We may rather wonder that this accident, which we call *stricture*, does not occur oftener, and that so delicate a part can ever escape injury. It is, indeed, certain that this affection is much more frequent than generally reported; for, according to so eminent an authority as Hunter, one may have it a long time unsuspecting; feeling the various attendant, remote, and constitutional troubles it brings, without adverting to the source they arise from.

However, strictures are of two kinds very different, the permanent and spasmodic; of which the spasmodic usually occurs suddenly, and draws immediately the attention of the sufferer. The irritation of the gonorrhœa incites the muscles of the urethra to contraction; which press so closely the sides of the passage together, at some one part more than the rest, that the urine cannot pass. Something of this sort, only of a minor degree, happens in every sharp attack in young persons, who, when they come to the last drops of urine, pass them with difficulty. But it most frequently occurs when the excitement of ardent liquor has been added to the complaint, with some exposure to cold: as when the patient sits drinking in company, and goes into the open air to empty the bladder. He is naturally alarmed that the water will not pass; that with repeated efforts he can only press out a few drops; while, in the meantime, the kidneys continue to act, and fill up the bladder, like a reservoir, to a greater height. Heat, flushing, and fever succeed, upon the imperative necessity of relieving the system.

In this it is that the spasmodic is distinguished from the *permanent stricture*, which supervenes only insidiously, and never ceases of a sudden, or without aid. All inflammation is accompanied by tumefaction; so that the sides of the urethra, thickened in gonorrhœa, press in, as I have stated, and narrow the tube. But should this inflammation run very high in some one part more than the rest, the lining of the tube is here still more thickened, and becomes dense and unyielding. This thickened part in time is hardly less firm than gristle, and presents a permanent obstacle to the urine. It is true, there is a small opening left; but perhaps not one-tenth of the full diameter. Not one only, but a number of these strictures may have their seats in the same urethra; and that illustrious surgeon I have already quoted saw half-a-dozen in one person.

Nothing is more desirable than that the patient should know by what symptoms he may discern the earliest approaches of an affection so distressing in its present effects when established, and so dangerous ultimately. "The earliest symptom," says Sir Astley Cooper, " is the retention of a few drops of urine in the uretha after the patient has made water; which drops soon escape, and slightly wet the linen." Now is the time for the patient to be on his guard, and arrest a further progress by a timely interference. " As the disease advances," continues Sir Astley, "the urine cannot be kept so long in the bladder, and most likely must be evacuated in the night."

This, however, is but a slight inconvenience in comparison of what follows; for the stream of water becomes gradually less, and is thrown spirally, like a corkscrew; at length it is diminished to a few drops, requiring much effort and force to expel them. This effort only hurts the diseased part more and more; which, finally closes in, and entirely stops the passage. It will happen, that, in this extremity, the urethra, over-distended, ulcerates, and the water forces a way downwards between the thighs, where it escapes with much injury. This not happening, the bladder inflames; other vital parts are involved; and

Fig. 15.*

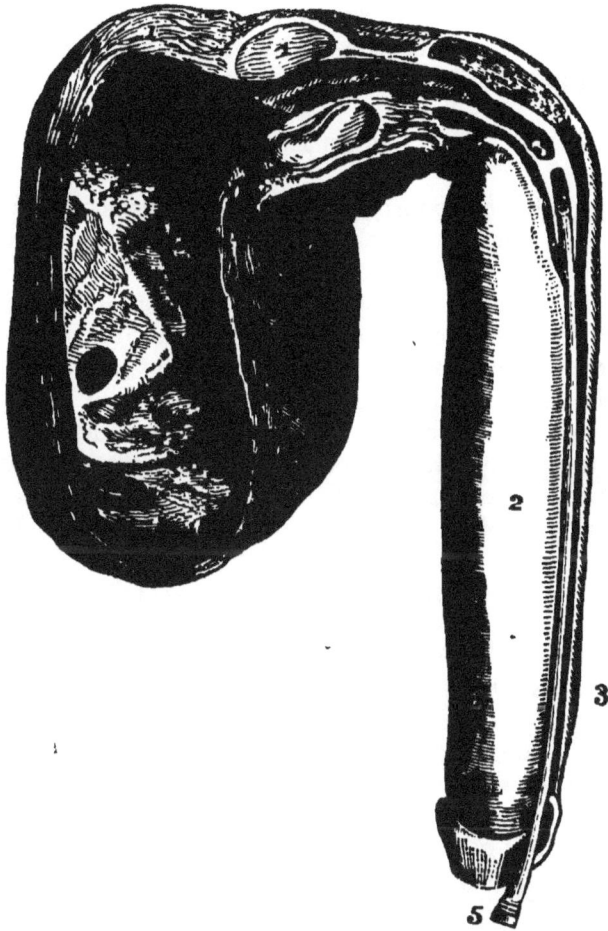

* The Bladder and Penis of a person who died of a mortification
of the bladder, in consequence of a stricture and stone in the
urethra.   In this plate, not only the stricture is represented, but
the thickened coats, and fasciculated inner surface of the bladder;
as also the small stone which acted as a valve, or plug ; besides
which, a Canula is introduced from the glans down to the stricture.

1, 1 The Bladder, cut open, showing its coats a little thickened ;
2 the Body of the Penis; 3, 3 the Corpus Spongiosum Urethræ,
cut open through its whole length, exposing the Urethra ; 4 the
Prostate Gland divided ; 5 a silver Canula introduced into the
Urethra; 6 points out the stricture, with the Stone resting above,
so as entirely to prevent the passage of the urine.

great suffering is put a stop to by death. I should remark, that this fatal termination is often very distant; for the disease is slow; and, as one of the latest authorities remarks, " a person, in consequence of inattention or ignorance, may for a long time be subject to stricture without paying any regard to it."

RHEUMATISM.

One effect of gonorrhœa must not be overlooked, which, although none more painful, may yet so far escape the sufferer, that he does not suspect the cause from which it comes. This is rheumatism, appearing usually only in one joint; but with so much the more intensity. The knee or wrist, for instance, may be the joint attacked; not without intolerable pain and anxiety. When unchecked, the rheumatism having run its course some weeks, on subsiding shows the violence of it in its remains. The joint continues through life immovable, which is that state termed *anchylosis*, from the Greek, ἀγχύλος, crooked.

GONORRHŒAL OPHTHALMIA.

The last affection I described occurs by what is styled a *metastasis;* the sudden translation of irritation to some remote organ. All parts of the body are, fortunately, not liable in gonorrhœa; but there is one very delicate part sometimes affected. However, not only may it be attacked by a metastasis, the gonorrhœal matter may by some chance get upon the eye, inflame it, and produce the same contagious fluid. The pain need scarcely be described, with the brick-red colour. According to the delicacy of the organ, is the rapidity of the disease, which, if not stopped before twelve or twenty-four hours, usually terminates in blindness. Let me give a humane caution: whoever has gonorrhœa, let him be on his guard not to touch the eye when he has been handling the penis.

CHAPTER II.

IN the treatment of gonorrhœa doctors are divided in their opinions, some preferring the irritant, others the soothing method. This last is the mode I have almost invariably followed. The symptoms should be at once checked, as soon as manifest, without waiting for any further developement; but only such prophylactics used as do not endanger worse consequences. The inflammatory symptoms should be reduced first, before any revulsive means are resorted to. It is then that the astringent gums are in their proper place, in graduated doses. The dragon's blood is, in many cases, an excellent astringent; in others, the Java pepper may be administered. The patient may drink orgeate, or syrup of capillaire. To lessen the scalding, put about twenty grains of bi-carbonate of potash, in a mixture with a drachm of the tartrate of soda and potash, first dissolved in a tumbler of warm water, and then mixed with a bottle of double soda water. However, this will not be powerful enough when the irritation is great, which must then be reduced by other means, according to the peculiarities of the case.

I will relate one case, which will answer the purpose of a hundred, to illustrate the complications of the disease. I find it in some old notes of mine. A gentleman of forty years of age, had a thin ichorus discharge from the penis, occasionally tinged with blood, with an obstruction to the flow of the urine, and pain in passing it. He had been for some time under treatment, when he was sent up to me from a provincial town, with a long statement of the means which had been attempted to relieve him, or, in medical phrase, the history of his case. The prevailing notion acted on was, that he had a stricture; for which bougies, catheters, and other means, had been used. At one time, stone in the bladder, with calculi in the urethra, was suspected, and an operation spoken of. The morning he called at my apartments, handing me his letter, I noticed some blotches on the forehead, which I concluded at once were syphilitic, before I had asked him any questions. I was therefore surpised to find his case described only as a gonorrhœa, with no other complication than stricture. I confess I had my doubts, and proceeded to examine him accordingly. The first thing that struck me was the peculiar character of the matter, which was neither the watery sanies of gleet, nor the rich fluid of gonorrhœa : it was more like genuine pus, with a slight sanguine discoloration. "I beg, doctor," said he "that you will not pass the catheter; I cannot endure the torture of it." I told him I had no intention of the sort; and pressed with my finger gently down a short way on the under suface of the penis, when I came to a slight ridge, or internal eleva· tion, and caused some little pain to the patient. I next took a drop of the discharge, and laid it on the slide of the microscope. I had now all the facts I required to come to a decision, and form a sufficient diagnosis: I concluded that the elevation I pressed was a chancre in the urethra; that the matter was the genuine secretion from it; that it was this which obstructed the urine, and was injured by the point of the catheter; and, finally, which left no doubt that it was the origin from which those secondary symptoms had come that I had. noticed on his forehead. On this supposition I treated him; put aside his gonorrhœa medicine, and cured him by those means adapted to the other form of disease, or syphilitic, which, as the reader will see presently, are entirely of another kind.

A great deal might be added here on this subject of treatment, which my limits necessarily exclude; for when the reader reflects that large volumes have been written on each separate form of these diseases, how is it possible to give full instructions in an abridgement concerning all of them? There is one case only which I will particularize, it is that of Spasmodic Stricture, which comes on so suddenly that there is no time for deliberation. As this matter has been very well described, together with the mode of treating it, by Sir Benjamin Brodie, I will here transcribe the passage, that the reader may have it done by an abler hand than mine:—

"A man who is otherwise healthy voids his urine one day in a full stream. On the following day, perhaps, he is exposed to cold and damp; or he dines out, and forgets, amid the company of his friends, the quantity of champagne, or punch, or other liquor containing a combination of alcohol with a vegetable acid, which he drinks. On the next morning he finds himself unable to void his urine. If you send him to bed, apply warmth, and give him Dover's powder; it is not improbable that in the course of a few hours the urine will begin to flow. After the lapse of a few hours you give him a draught of infusion of senna and sulphate of magnesia, and when this has acted on his bowels, he makes water in a full stream."

So far Sir Benjamin. I will add, that a hip-bath of hot water, just before taking the Dover's powder, will much assist in relaxing the Stricture, and enabling the urine to escape.

SYPHILIS.

CHAPTER I.

SINCE the researches of Carmichael and others, it is now generally allowed that there are at least *four* varieties of chancre, with four varieties of secondary symptoms, each peculiar to, and succeeding only on, its own primary sore : in such a manner that, the surgeon inspecting the chancre, he may predict what particular secondary symptoms may follow. The names which have been given to these varieties are not taken from the primary sore, but from the secondary symptom, as the most serious and constitutional part of the malady. These four forms of the venereal disease are the *papular*, the *pustular*, the *phagedenic*, and the *scaly*.

SECTION I.

FIRST CLASS OF SYPHILITIC DISEASES.

THE PAPULAR FORM—PRIMARY SYMPTOMS.

THIS is by much the most frequent variety met with, not only, I may say, in this country, but in others where I have travelled, as it appears in their several hospitals. After impure connection, a period of from three to eight days usually elapses before any change may be observed in the organ; not that the patient is safe even for weeks after, though nothing has appeared. The earliest sign is a slight, and perhaps not disagreeable, itching ; with this a slight redness comes on, and subsequently a more regular areola or circle a little inflamed. In the centre of this areola rises a minute pimple or vesicle, filled with a transparent fluid. After a few days, there is regular pus in the chancre, which dries, and forms a scab, while ulceration is going on beneath. The scab separates, and the ulcer is then discovered excavated, with a thin ichor exuding from it.

SECONDARY SYMPTOMS OF THE PAPULAR FORM.

WE can tell, in the small-pox, to a day when to look for the eruption, from the date of inoculation. Not so in these complaints; for as the appearance of the sore may be deferred for months after contamination, so the constitutional effects have no certain period, from weeks to years.

However, with regard to the present form of the disease, possibly as early as the third week, the patient complains of lassitude, with chill and nausea. These are the usual precursors of fever, which continues but a few hours before an eruption shows, first on the forehead and breast, afterwards over the body and limbs. This eruption consists in a number of small elevations, hard to the touch, and of a deep red colour; they are but so many separate ulcers, which in a few days are covered with incrustations, that come off in the form of scales.

Simultaneously with this eruption, there is a rash or efflorescence in the throat, with a swelling of the tonsils, which give pain in swallowing. The glands of the neck may be enlarged, and prove very troublesome to those with a tendency to scrofula. It is very common, also, for pains to begin at the joints, like those of rheumatism; particularly in the hip, knee, and shoulder; getting worse at night, and heightening every other inconvenience by the loss of sleep.

There is a secondary affection, which I have noticed not only after this, but succeeding to every other form of primary ulcer. It consists of superficial ulcers inside the lips and cheeks, at the angles of the mouth, and of fissures on the sides of the tongue. This affection sometimes occurs very late; often when every other symptom has disappeared: as if to illustrate the malignant tenacity of the disease, and put the practitioner on his guard against every contingency.

E

SECTION II.

SECOND CLASS OF SYPHILITIC DISEASES.

THE PUSTULAR FORM.—PRIMARY SYMPTOMS.

As the last is the most common, this is the most rare, of the venereal forms; yet not obviously different from the preceding at the outset. It is not until two or three weeks that the margin of the chancre rises above the surface, while it is undermined beneath, making it slow to heal, and being the chief character that discriminates it. As that I have already described is the superficial, this may be styled the ulcer with elevated edges.

SECONDARY SYMPTOMS OF THE PUSTULAR FORM.

The earliest alarm is given, as usual, by the accession of fever; to which succeeds an irritation of the skin, especially of the forehead, giving rise to pustules, that may discover subsequently all over the body. These pustules soon become so many ulcers, especially troublesome in the head at the roots of the hair. There is an offensive odour, and altogether the symptoms are much more severe than the preceding; in other respects not so essentially different from it. The throat is sore and dry, with little white flakes that may be seen on it, which the patient makes repeated efforts to swallow, or expel by coughing: a most distressing annoyance, which interrupts the sleep, and provokes to continual irritation of mind, not less than body.

The joints, also, are not simply the seat of rheumatism, but are much inflamed and distended.

SECTION III.

THIRD CLASS OF SYPHILITIC DISEASES.

THE PHAGEDENIC FORM.—PRIMARY SYMPTOMS.

Those other primary chancres I have described, are soon bounded; the immediate injury is, in comparison, trifling, only except as it fires the train to more distant ravages. The phagedenic sore, on the contrary, spreads from the beginning, seeming to consume where it touches, as its name implies, which is from the Greek, φαγω, to eat. It has, therefore, a corroded appearance, with jagged edges, which are undermined, and the outline irregular. The process by which it spreads is from a slough, or scum, of a greyish cast, which comes off every few days, with a little bleeding. It is this which gives the sore an unclean surface, unlike the other chancres, from which the matter may be washed. As each slough separates in order, the excavation widens proportionally, so as to cause the most serious mutilation, if not stopped by appropriate remedies. Carmichael has seen the entire penis go, with the scrotum, " leaving the testicles perfectly bare."

SECONDARY SYMPTOMS OF THE PHAGEDENIC FORM.

WITH so much severity of the primary affection, it is natural to expect the re-appearance in the constitution with corresponding severity. The eruption rises in pustules, from which the matter exudes, and drying, forms elevated crusts, often of a conical outline, especially on the face. This is by far the most unsightly, painful, and disgusting form of eruption ; the linen stained with the ichor, and the odour foul. And even when on the decline, the skin continues in scars, and not a little discoloured.

" The ulcers of the throat," says Carmichael, " which attend the phagedenic form of venereal, display, to the full extent, the virulence of this disease. There is no part of the fauces, but more particularly in front of the bodies of the vertebræ, that is not liable to be assailed; and whereever it commences, it spreads with a frightful rapidity to all the other parts ; so that it is not unusual to see the velum, uvula, tonsils, and back of the pharynx, engaged in one foul ulceration, extending upwards into the nares, and downwards into the œsophagus, destroying the epiglottis, and penetrating to the larynx. The affection of this last organ produces a train of the most distressing symptoms, which sooner or later, if not met by the timely operation of tracheotomy, causes the death of the patient." Having completed his lecture, Mr. Carmichael shewed to the students a number of preparations, to illustrate his remarks. " In this preparation," said he, " exemplifying extensive ulceration of the pharynx, the lingual artery gave way, and the patient died of hæmorrhage, before assistance could be had to secure the vessel. In this other preparation, you observe caries of the bodies of two or three of the cervical vertebræ, the consequence of phagedenic ulcers of the pharynx."

SECTION IV.

FOURTH CLASS OF SYPHILITIC DISEASES.

SCALY.—PRIMARY SYMPTÔMS.

In this form of disease there appears in the second or third week in the base of the sore an unusual thickening or condensation, which gives the idea of a piece of cartilage beneath the skin.

SECONDARY SYMPTOMS OF THE SCALY FORM.

Not less tardy of advance than in the primary, are the secondary symptoms; ushered in with scarce any fever, and coming at first with a general efflorescence of the skin, but more observable on the forehead, neck, and breast. It separates, after a while, into patches, copper-coloured, which subsequently are covered with scales, that give a name to this particular form. The throat is attacked, as in the other varieties; but with a violence so little to be measured by the pain, that, as Carmichael remarks, the surgeon is often the first to make the patient aware of what is going on, though the tonsils are already excavated. The ulcer is foul, with a thick slough, not easily washed away.

Externally, the glands of the neck are swelled, and break in what may be denominated consecutive bubos.

CHAPTER II.

BUBO.

There is one occurrence requiring a more particular consideration, following close on the chancre, no matter of what variety, or coming as the other declines. This is bubo; a swelling that rises in the groin; making a sort of intermediary affection, between those more accurately called primary and secondary.

The earliest notice of a bubo is a stiffness in the groin, and then tenderness immediately over the place. A round elevation succeeds, not painful until later, when the tumefaction extends, and the surface is red. It rises more to a point by degrees; there is throbbing; and, on slight pressure, matter may be felt underneath. There is much burning and fever, until the abscess opens of its own accord, by ulceration, with a copious discharge.

Only one occurrence, consecutive usually on chancre, remains unmentioned; it is a species of vegetation on the penis, spreading large, and with a cauliflower excrescence. A number of granules rise upon one another, like warts; and often increase to a size much larger than the organ itself. They are very tender, and bleed easily. They sometimes encroach upon the opening of the urethra; and even appear in clusters round the anus, to the great inconvenience of the patient.

CHAPTER III.

TREATMENT OF CHANCRE, BUBO, AND SECONDARY SYPHILIS.

I now come to explain the method by which a venereal primary ulcer, with those effects called secondary we have seen succeeding to it, may be met by remedies of art; a subject which the closing remarks of the last chapter have naturally introduced. Were we to consult authors on this occasion, we should find them, as in gonorrhœa, differing in all their opinions, and ready to confound at every issue the zealous inquirer after truth. Whatever of this contradiction does not come from the motives I have already assigned, such as the vanity of opinion, and the interest of singularity, may be ascribed to the neglect of circumstances, peculiar in every case. For when a surgeon asserts that he has cured hundreds by a special remedy, which another declares has entirely failed in his hands, what are we to infer except some vital variety in the cases themselves, the stage at which the remedy was used, or some accidents entirely overlooked? unless we reject one or both statements altogether, which the claims of respectability, not less than the sincerity of science, will not allow of.

I have sometimes said to patients bringing with them the prescriptions which had failed, I can cure you by those very medicines; which I have done by altering the quantity, the time of use, or the mode of combination; for frequently ingredients are unskilfully combined, and decompose one another in the manipulation. The patient had perhaps been kept on too low a diet, confined unnecessarily to the house, denied exercise, while larger doses were resorted to: a species of blunder which is obvious enough; when, in proportion as the constitution was more weak, the remedy was stronger.

I remember a ludicrous accident, related to me by a patient, which will very well illustrate how the same course of treatment may be inverted with advantage, without other change: by mistake of direction, he swallowed the injection, and injected the dose: which, to his own great amazement, not less than his doctor's, effectually restored him.

As the basis of a philosophic practice, the first consideration is the variety of disease itself, and that, under the same name, several forms may not be confounded. It is for this purpose that I have classified symptoms, as the reader has seen, under four different aspects, drawn as disease itself had previously laid the outlines. The superficial primary ulcer is one; that with elevated margins another; different from each is the phagedenic sore; while the Hunterian chancre, with its cartilaginous base, is the most peculiar of all. These respectively demand a separate management, and frequently what is beneficial to one of them, will be quite the reverse with another. The same may be said of secondary symptoms; and nothing, therefore, is more essential than an exact definition and perfect comprehension of each.

As it is not always that I have the management of chancres from the first, which often pass through other hands previously, I will mention the treatment proper through the different stages through which they run. When there is inflammation, it is necessary, as a preliminary measure, to reduce it; this is done by the antiphlogistic regimen, and cold applications, with oiled silk. On the other hand, when there is atony, I recommend a more generous diet, tonics, and other restoratives. There is always an especial regard to be paid to the constitution, some patients requiring, from this cause, an entirely opposite course to others. Part of my advice is negative ; I reject all rancid ointments, and irritating washes, often the only obstacle to a cure. When the sore is very painful, the vinum opii may be used as an external application. In other cases, the aromatic wine* of the French codex may be medicated with tannin, especially when the ulcer secretes copiously, and applied with lint. The phagedenic chancre should be treated with extract of belladonna, dissolved in water, and put on in the same way with lint.

One rule, however, is universal : venereal sores of whatever kind should be managed gently; and I am here happy to quote the expressions of a modern writer, entirely in favour of the practice I have found successful for so many years : "The more mildly primary ulcers are treated locally, the less likely are they to be followed by those appalling complications which sometimes accompany them, such as rapid ulceration, sloughing, or disorganization of the penis and scrotum, which used to be common under the old treatment of stimulating mercurial applications during the first days of chancres."

All means, such as above suggested, are, of course, only of inferior utility, compared to such medicines as are known to be absolutely antisyphilitic. The common plan is mercury, which, although denounced in the passage just quoted, is still recommended. Why should this be so ? perhaps suggests itself to the reader ; and if this mineral is so destructive, why is it so obstinately persevered in ? This is a question I cannot easily resolve to one who has not considered the force of prejudice, and how the greatest abuses are continued long after they have been exposed. But there is a better reason than the force of example or prepossession ; it is the force of ignorance ; when those who want the skill and the address of a nicer machinery, have recourse to the first rude method, which will kill or cure, without more trouble.

There is one cogent reason why the rational, as it is called, should be preferred to the mercurial, treatment :

* The formula for the aromatic wine is as follows : Take of aromatic herbs four ounces, viz., rosemary, rue, sage, hyssop, lavender, absinthium, origanum, thyme, laurel leaves, the flower of red rose, chamomile, melilotum, and elder ; red wine two pints : digest for eight days.

the patient may go abroad without danger from cold or
wet. It is all very well for practitioners to instruct a
patient to confine himself to his room; when, most likely,
his avocations will not permit him, when business calls, or
when he must shut out suspicion. Nothing, in reality, is
more beneficial than the open air to persons affected with
syphilis, and many sores will grow worse, even with good
treatment, in a close room. It is necessary, indeed, not
to irritate the part, or excite the circulation ; but exposure
to pure air, which dare not be ventured in the *mercurial
complaint*, in the *venereal* has never been suspected of injury.

One of those medicines, which may be substituted for
mercury, is a solution of antimony, with the tincture of
opium, taken in minute doses, so as to keep up an insensible diaphoresis, and interrupt specific action, but without
nausea. No sooner, however, has the ulcer, whether
primary or secondary, put on a certain languid appearance,
than the antimony should be discontinued. The iodide of
potassium is another of these remedies ; but should be
administered in moderate doses, without saturating the
system, or disordering the stomach. It may promote
absorption too much, in the phagedenic ulcer especially,
which it is necessary to observe. Iodine may be administered alone, which many regard as a potent antisyphilitic,
independent of its general virtues. The muriate of gold
is more beneficial as an external application ; but may be
substituted advantageously for the bichloride of mercury,
under the same forms, and in the same doses. There is a
substance called nitrate of ammonia little known to the
Faculty, exceedingly efficacious, especially in the secondary
forms of syphilis. The proportion and doses of all these
substances will depend, of course, on the age, sex, constitution, temperament, stage, and state of the disease.

Local affections must be treated as they appear. Incipient
bubo may be met by a solution of muriate of ammonia,
and extract of belladonna, in rose water ; or, the iodide of
lead, with belladonna. For sore throat, acidulated
gargles, when there is only inflammation ; but chloride of
calcis when there are ulcers. Fumigations are especially
applicable when the nose is affected. For nodes, steam is
an admirable auxiliary.

These remedies are all very useful, each in its place, and in experienced hands. With others, they are but edged tools, that may do more injury than benefit. I am far from saying that I have enumerated these in deference altogether to high authority ; yet I must declare that my own practice is in many respects peculiar ; which, I suppose, is no more than every practitioner will say, who chooses to observe and reflect for himself. The volume of nature is open to all, and the characters in the universal language, though few will be at the pains, or have the perseverance, to study it.

The chief respect in which I rely on my own experience, is in the preference of vegetable to mineral medicines; for, having rejected mercury, I was more and more, and by degrees, diverted towards agents of the opposite kind. The great, or physiological, object of all remedies, is to stop that peculiar action, local or general, which constitutes disease : one class, which is that I reject, in lessening the morbid action, lowers, at the same time, all the natural operations of the system also, or substitutes another, and perhaps a worse, species in their place : the other class of remedies, or that I recommend, supports the general functions, while it alters the specific influence ; and lessens irritability, but increases vital power. The supported energy of the constitution, is, in the one case, made to overcome the disease ; while, from causing a greater weakness, in the other, the remedy itself acts, indirectly, as an accessory to it. To impart such minute instruction as to apply in every phase of the symptoms and complication of collateral circumstances, is what has never yet been attempted. If the subject were more limited, one might exhaust it, and write a larger book ; but here the topics are so various, and the exceptions so many, that there is no possibility of including them all : general principles are all that can be laid down; and the extraordinary extent of the materials is the cause, in an unusual way, of the brevity of the treatise.

To any member of the Faculty who may desire it, I shall always be ready to explain my peculiar views, or give suggestions for any particular or difficult case. To patients, while I am always attentive to the great purpose, which is the cure, I frequently explain the methods by which I

effect it. Those who may wish to consult me, are requested
to do so at a stage as early as possible of the disease, and
soon after exposure, when every manifestation may be
anticipated. Those at a distance may describe the symptoms (for which any plain language will answer); and
enclose a small piece of rag, stained with a drop of the
matter, whether of chancre, gleet, or gonorrhœa. He
who has studied these pages will know my object, which
is to submit the secretion to the microscope.

INSTRUCTIONS TO PATIENTS.

I DESIRE in this place candidly to explain to the patient that, on account of the great variety of constitutions and disease, it is in far the greater number of cases utterly impracticable to give instructions sufficiently ample to enable him to conduct his own cure. He may understand this more fully when I inform him that large volumes have been written on the treatment of one accident alone of Gonorrhœa, as Stricture for instance. How, then, is it possible to give in a small book minute instructions by which a person unacquainted with the science of medicine may prescribe for that particular form of complaint which by chance his disease may have assumed? In former editions of this work I endeavoured to compress into a few pages the more usual particulars of treatment in the more common cases, with a few what I supposed plain prescriptions; but, notwithstanding that I made the matter as clear as I could, I was mortified to find that in many instances patients misapplied my rules, made mistakes which led to mischievous consequences, and brought into disrepute directions which, had they been judiciously followed, would rather have produced credit for myself and my book. Under these circumstances I have thought it best to omit altogether those special instructions, which to understand and apply in an efficient manner one must be a physician himself.

However, some cautions omitted in the previous part of this work may here be inserted, such as those relating to the early appearance of disease. Be careful after the appearance of excoriation on the penis, or of a pimple, however small, coming on after a suspicious connection, but especially when a drop of discharge has been seen at the opening, to wash and cover the part with a rag, that by contact the disease may not be communicated to some other part; and never let a towel be used, lest in wiping the hands, and afterwards by accident touching the nose or eyes, those delicate parts should be inoculated also. Avoid all quack nostrums, which at best are simply inefficient, and lucky for the patient if they are not hurtful. It is always best to apply to some properly educated physician, and best to one who has made an especial study of this class of disease, the earlier also the better; the disease instantly spreads its roots, and in no instance dies out of itself.

One interview at least is desirable, in most cases, as affording facilities of tracing symptoms, much less easily collected by letter.

Finally, I have only to make known that I give my exclusive attention, as I have done for the last thirty-six years, to that class of effections herein described; for which purpose I am in **Daily Attendance**, and may be consulted personally from **Eight in the Morning** until **Nine at Night**.

WILLIAM WATSON, M. D.,

New York

www.ingramcontent.com/pod-product-compliance
Lightning Source LLC
Chambersburg PA
CBHW021947190326
41519CB00009B/1170